On Site

1921–71

ALAN JENKINS

HEINEMANN : LONDON

William Heinemann Ltd
15 Queen St, Mayfair, London W1X 8BE
LONDON MELBOURNE TORONTO
JOHANNESBURG AUCKLAND

© Taylor Woodrow Services Ltd, 1971
First published 1971
434 90890 8

Printed in Great Britain by
Cox and Wyman Ltd,
London, Fakenham and Reading

Contents

List of Plates

INTRODUCTION

This is the Taylor Woodrow story of fifty years' loyal endeavour and progress. It was my good fortune to start the company, and therefore as you will appreciate, it is a great joy for me to write this foreword to our story. Please accept that the book can only be a brief digest and that it will be impossible to mention the names of everyone who has contributed to our growth, but may I say to all team members who have made possible this progress and the establishment of our reputation in the forefront of our industry today, sincere thanks and congratulations. May I also take this opportunity of thanking the many clients throughout the world who have placed their orders with us and the large number of architects, consulting engineers, quantity surveyors, and other professional people with whom we have been associated.

All those young people in our team in the great challenge and adventure ahead are going to carry on in the future, adding lustre to our reputation and expanding our business and profits. They will, I feel sure, remember the policy on which Taylor Woodrow was founded and has continued throughout the whole of its history, of being fair and giving a square deal to all with whom we come into contact, within the team, to our shareholders, our suppliers, our clients, and to the public at large.

This is not the end of the story, but only the beginning of the second fifty years, with an era of exciting potential ahead.

Having read the manuscript, I would like to congratulate Mr Alan Jenkins on succeeding in the small compass of a book in giving such a good, broad picture of the Taylor Woodrow team and our activities.

FRANK TAYLOR

Acknowledgments

THE author would like to thank the following people for their help in writing this book: Mr and Mrs Frank Taylor, for much personal kindness and many of the stories in the earlier chapters; Nat Fletcher and Eric Sadler, of Taylor Woodrow's Publicity Department, for tireless research assistance, for their long memories and their sound advice; the several Publicity Department members who have been deputy editors of *Taywood News*, including Elizabeth Stilling, Allan White, and currently Guy Priest; countless team members, at home and abroad, for the friendly patience with which they answered my layman's questions, and for sparing me so much of their time; those who read the manuscript and corrected many errors; and to the team as a whole for allowing me the honour of being a temporary Member.

My thanks are also due to Mr Ian M. Leslie, O.B.E., editor of *Building*, for permission to consult fifty years of files, and to Frank O'Shanohun Associates for information and advice. The following publications were among many consulted:

Punch, 1921.

Flames Over Britain, by Sir Donald Banks (Sampson Low, Marston & Co. Ltd).

Portrait of Churchill, by Guy Eden (Hutchinson & Co. Ltd).

Crusade in Europe, by Dwight D. Eisenhower (William Heinemann Ltd).

A Dictionary of Civil Engineering, by John S. Scott (Penguin Books, 1958).

Journal of the Royal Society of Arts, July 1963.

Metropolitan Cathedral of Christ the King, by Sir Frederick Gibberd, C.B.E., A.R.A.

A Cathedral for Our Time, by Patrick O'Donovan.

Survey of the Building Industry – 1962 Onwards (*The Builder* & O. W. Roskill, 1964).

The Property Boom, by Oliver Marriott (Pan Books Ltd, Hamish Hamilton, 1967).

Singapore 1970 (Ministry of Culture, Singapore).

'Happiest Spot in Nutland', by Noel Monks (*Daily Mail*, December 20 1949).

'The Tough World of Frank Taylor' (*The Director*, March 1964).

The Stock Exchange Gazette, November 26 1965.

'Hinkley Point/Sizewell/Wylfa Head: A Logical Development' (*Nuclear Engineering*, 1965).

Business and Industry (B.B.C., External Service, November 15 1967).

Silent and Vibration-free Sheet Pile Driving, by Taylor Woodrow's E. W. M. Page & W. Semple (Institution of Civil Engineers, 1969).

The Economist, July 5 1969.

The Guardian, July 9 1969.

'The Challenge of the Seventies', by Russell Jones (*Contractors' Plant Review*, November 1969).

Press & Journal, Aberdeen (January 12 1970).

Business Administration (February 1970).

New York Times (March 14 1970).

Investors Guardian (May 22 1970).

'Hartlepool AGR', by various authors (*Nuclear Engineering International*).

Author's Note

The hero of this book is a world-wide group of companies. For the reader's convenience, and to avoid interrupting the narrative with parenthetic explanations or footnotes, I shall try to summarize its structure as briefly as I can.

At the top, in an elegant red-brick house in Park Street, Mayfair, is Taylor Woodrow Ltd, the holding company of the 'Group'. At Southall, Middlesex, is the headquarters of the principal subsidiary of the group, Taylor Woodrow Construction Ltd, which designs and constructs building, civil, and mechanical engineering works of all kinds, and has regional companies at Stafford and Darlington, and a Scottish office; and includes sections such as the atomic power department and the mechanical and electrical division.

Taylor Woodrow International Ltd., based at Western House, Ealing, W.5, controls and carries out building and civil engineering work in overseas territories where there is no local Taylor Woodrow company. It has regional offices in Guyana, East and West Australia, Gibraltar, Singapore, and Malaysia. Other overseas companies based at Western House have offices in such busy areas as West Africa. At Western House also are Taylor Woodrow Homes, developing housing sites throughout Britain, and Myton Ltd, building and civil engineering contractors, acquired in 1955 and with many major projects to their credit. Nearby West Africa House is the headquarters of Taylor Woodrow-Anglian Ltd, which specializes in factory-made housing using a cross-wall form of construction with large concrete panels, based for multi-storey homes on the Danish system of Larsen & Nielsen and, for low-rise homes, on their own designs. Taylor Woodrow-Anglian's two factories (at Rushden, Northants, for the South and Sunderland, Co. Durham, for the North) have a total potential output of components for 4,000 homes a year.

From Welbeck Street, London, bases, Taylor Woodrow (Arcon) and Taylor Woodrow (Building Exports) Ltd supply Arcon structures to U.K. and overseas markets respectively (there is an Arcon company in Singapore too), and Taylor Woodrow Industrial Estates specializes in projects such as the Dixon Blazes estate in Glasgow.

Taylor Woodrow have their own Plant Company, for both mechanical and non-mechanical plant and vehicles, which are supplied to companies in the group, especially Taylor Woodrow Construction. They also have Greenham (Plant Hire) Ltd for hiring to anybody, and the Greenham group includes a wholesale electrical company; workshops for overhauls; a chain of ready-mixed concrete plants; a sand and ballast company; and companies for tools and tyres and dredging aggregates from the sea. Although Taylor Woodrow (East Africa) closed down in 1956, Greenham are still there, for supplying civil engineering plant, air-conditioning, and lifts. There are also Greenham companies in Ghana, Nigeria, and Transvaal.

Taylor Woodrow Property Co. Ltd., still a young company, is likely to point one of the ways to the future, as will be seen: they have regional offices in San Francisco and Australia. There are, besides, specialist companies such as Jonathan James, with its subsidiary Rainham Fixings, who make floor screeding, suspended ceiling work, plaster, and mouldings which are in demand for, among other things, restoring old houses; Terresearch Ltd., a building laboratory which studies ground exploration, soil mechanics, piling, and foundations; Taybol Ltd., jointly owned with a Netherlands interest, set up to do dredging all over the world; James Sim, builders and civil engineering contractors, of Kirkintilloch; and F. Wykes Engineering, of Poole, who make precision components for aircraft and also jigs and tools.

The group is in Ghana, Nigeria, Sierra Leone; the U.S.A., where it has an interest in the Blitman Construction Corporation; the Bahamas, Belgium, Mallorca, Malta, Denmark, Milan, and Paris. In Canada there are eleven subsidiaries and associated companies; in Australia another eleven, some in partnership with local companies; and one in New Zealand, also in local partnership.

Swiftplan Ltd (formerly known as Builders Supply Co.), based at Southall, with also Darlington and Kirkintilloch offices, makes industrialized buildings and demountable partitions, paying special attention to noise reduction: in 1960, this company built a pub in a day – the Rising Sun at Clanfield, Hants; and in 1969 it put up an office block in seventeen weeks.

I

Holes in the Ground

THE word 'Team' is going to be used rather a lot in this book. What do Taylor Woodrow people mean by it? Who is in it? Well, you could say it consists of Frank Taylor, London; Peter Tabert, civil engineer, Hong Kong; Roy Wykes, Canada; Alhaji Hassan Yusuf, chief accountant, Nigeria; Dennis Warne, foreman carpenter, Ipswich; Takir Quarishi, quantity surveyor, Kabul; David McIntosh, labourer, Invergordon; Denis Ngawhare, crane driver, New Zealand. . . .

They stretch across the world, these Taylor Woodrow people; about 25,000 of them, all belonging to what City editors call a 'a £40m group' with a turnover of well over £100 million (and a pre-tax profit of £4 million), and over 100 companies, each with special skills or spheres of operation working in the five continents. Sometimes a new company or consortium is formed for a particular project; sometimes, when a project is completed in a certain country, there appears to be no further work immediately forthcoming in that area, and a company will be wound up.

Often whole families move about the world from site to site. Take Mrs Mary Gray, receptionist at the Taywood-Wrightson consortium's Invergordon construction camp, where they have been building a £37 million smelter for British Aluminium. She and her husband William, a building supervisor with Taylor Woodrow Construction Ltd., arrived

at Invergordon in the spring of 1969 – from the new Hong
Kong ocean terminal via Euston Station, both Taylor
Woodrow contracts. Before that, they worked at Sizewell,
Hinkley Point, and Calder Hall nuclear power stations.

Also at Invergordon were the Hodgsons, a family of six,
five of them working on the smelter site. Father, John
Hodgson, is plant foreman; his wife is chargehand office
cleaner; daughter Sandra is a typist in the site office; son
David is a concrete engineer; and his brother Edward an
apprentice fitter. They all live in a caravan on a farm at Alness
near by; just as they have lived at Fawley, Calder Hall,
Hinkley Point, Sizewell, and other sites during the past
fifteen years. The camp hostel is supervised by Mrs O'Keefe
of Reading, whose husband is a general foreman. . . .

And on the London Airport motor transport base, in 1966,
were seven members of the same Indian family, the Singhs, as
bricklayers and general labourers.

We cannot mention everyone. We can only select a handful of people who are typical enough to tell our story. It
would be possible to write a book as long as *War and Peace* on
this international group; there is a book in each subsidiary
company.

As we go along, we shall talk of dams and power stations,
railways and cathedrals, tower blocks of offices, quarries,
aggregates, bulldozers, concrete, and little houses where,
what Frank Taylor calls, 'salt of the earth' people live. People:
we shall try never to lose sight of the fact that every operation,
even the programming of a computer, is carried out by a
human being, and it is these people who are actors in the adventure story we are trying to tell.

The author, it will soon become clear, is not a civil engineer
or even a bricklayer; he is easily troubled by words like slip-
form, spud leg, traxcavator, and muck-skip; and he tends to
find a *Troubleshooters* kind of drama in events that sober

). 1916. Sarah Taylor and her eleven-year-old son, Frank.

1(b). 1921. Frank Taylor aged sixteen with his first lorry.

1(c). The first pair of houses in Central Drive, Blackpool.

2(a). The Thirties. Southern venture – the start at Grange Park, Hayes.

2(b). Part of the Cranford Park Estate, Hayes.

3(a). A very important visitor tours the Mulberry Harbour construction sites

3(b). Post-war housing. An estate of Arcon prefabricated bungalows at Great Yarmouth.

4. Atlantic House, Holborn. London's first major post-war office development.

engineers take for granted. And yet do they? There is always a flashpoint at which a Taylor Woodrow man's eyes begin to gleam with technological excitement as he tells you about his job and what happened at the rutile mine the day the rains came...

Companies, and groups of companies, grow partly by chance and partly by design. The Taylor Woodrow group does not 'take over'; it 'merges' or 'acquires an interest'. The result is not so much a conglomerate, more a flexible chain-mail of convenient-sized links, of teams within teams, which describes itself as 'Taylor Woodrow, the world-wide team of engineers, constructors, and developers'. Note the order: it is significant for the future. Housing, which is how nearly all great construction groups begin, now forms only five per cent of the group's output.

How does a group like this get new business; and having got it, how does it plan to carry it out? New business, or 'work procurement', is the task of contract promotion or sales staff. The best new business often comes through past performance.

When a contract heaves in sight, whether it be a competitive tender, a negotiated contract or a 'turnkey' offer, the work goes first to the estimating department, where planning and pricing are correlated (nowadays with a computer's help). In the case of a turnkey offer, a project engineer is appointed to integrate design, estimating, and budgeting from start to finish. Sometimes there are very complicated financial arrangements: 'Thank heaven for our bankers!' Frank Taylor says. 'For nobody knows what contractors owe to their bankers.'

As soon as a contract is awarded, it becomes the responsibility of a production director, usually known as the 'director-in-charge', who appoints a project manager who will actually see to the execution of the work on site. He will

B

use the organized services of the company as far as possible; but if there is any hitch or supply problem, he has full powers to act on his own initiative.

Before any work or expenditure can begin, there has to be a numbered Building Order, signed by a director. This describes the work, where it is, and likely contract conditions; lists of the senior staff on site and those at H.Q. who will have to visit the site. Generally there is a contracts manager assisting the director in his liaison with the site. Every site is different: what are the physical conditions? Is there a local labour force? Will certain personalities get on with each other? Shall we have to lay on coaches to get the men to work from the nearest town or village, or is the site so remote that we shall need to build a whole temporary township, complete with shops, offices, clubs, a bar, a church?

Back at H.Q. there are service departments to help the project manager in the front line: personnel, buying, plant, quantity surveying, accountancy, safety co-ordination, legal. A nucleus of staff from the design department may be sent out to live on site. Research laboratories (again helped by computer) can in a few hours simulate conditions and produce data which would otherwise take years.

Everything depends on pre-planning and advance information: for instance, a whole contract can be delayed – and lose money – if the team on site finds itself up against a kind of rock which (through lack of information, or misinformation) wasn't mentioned in the initial research. That first 'recce' must investigate access to the site: will we need temporary roads? Is the whole place a sea of mud at certain times of the year? How the hell do we get at it? (The search for nickel in Indonesia must have posed this question hundreds of times.)

Work can begin only when all these preliminaries have been done. The site headquarters spring to life with briefing sessions on the spot, day and night telephone calls and telexes to and from head office, client consultants, suppliers both bulk and local; and much midnight electricity is burned at Southall, Western Avenue, and elsewhere.

The site is beginning to resemble 'a battlefield without blood': Frank Taylor described it in his lecture to the Royal Society of Arts in March 1963:

'Now things start humming . . . our construction work is done in the "great outdoors" where the field is our workshop floor and the sky our roof. Machines get into position, start excavating the site, concrete mixers are set up, transport starts bringing in temporary buildings, scaffolding, building materials; cement silos are erected, and there is a real hive and hum of activity. Engineers with their assistants and chainmen with their dumpy-levels and theodolites are carefully setting out the buildings, giving lines and levels; the project manager has a thousand and one things to do, briefing his general foreman, trades foreman, mechanical people, checkers. . . . It is still a thrilling moment when we start each contract and cut the first sod – I am glad to say it is always done these days with a big excavating machine.'

Most construction begins with a hole in the ground. Say, a ten-acre hole like the one that used to be Victoria Station, Nottingham, with platforms, services, tracks. (Seventy-one years ago it was a hole, too, ready for the building of the station.) In 1968 Taylor Woodrow's Midland construction company started filling that hole with £8 million worth of 'sophisticated multi-amenity complex' which will make it a residential area again, as it was a century ago. The scheme contains one of the largest covered and climate-controlled shopping centres in Europe, with department stores, smaller shops, a market, bus station, office blocks, entertainments, an underground car park, and 464 homes in multi-storey blocks.

Over six million bricks (about twenty for every inhabitant

of Nottingham) will have been consumed on completion; 125,000 cubic yards of concrete brought in 20,000 lorry-loads; and 10,000 tons (the weight of a handy ship) of rod reinforcement. About 500 men will have turned the liquid concrete, bricks, steel bars, acres of asphalt, plaster, finishing materials, windows, floors, timber, pipes, and wiring into the five-level, quarter-mile long Victoria Centre.

Nottingham people are fond of the old Clock Tower, and so, when the Duchess of Kent visited the site on October 14 1969, she unveiled a commemorative stone bearing these words: 'The Clock Tower remains as part of the former Victoria Station, which was opened on 30 May 1900, the anniversary of H.M. Queen Victoria's birthday.'

This was a pretty civilized 'hole in the ground'. In Sierra Leone, things could not be more different. From 1964 to 1966 Taylor Woodrow had several holes in the ground 160 miles from Freetown on the rutile mining contract. The team on the job tended to refer to Rutile as a place, not a substance; in fact, rutile is titanium dioxide, used to make white paint and high-temperature steel.

Here the task, apart from getting at the stuff, was to make life tolerable, even pleasant, for the team. Grahame Smith, project manager, wrote a report on Rutile which has made a major contribution to thoughts on camp-planning. 'Temporary Works on Remote Construction Sites' runs to about 5,000 words with maps, diagrams, and appendixes. Grahame Smith, who joined Taylor Woodrow straight from college, served his apprenticeship on opencast coal-mining in Derbyshire – 'a good, dirty introduction to civil engineering', he says. He knows the importance of human relations on a tropical site, latitude 8°N. – that there are some men (and some men's wives) who can't stand it, and that you must pick your team with extreme care.

From June to November, in Sierra Leone, it rains – about

seventy inches. The nearest open water was mangrove swamps and creeks. All supplies came from Freetown by road, the last fifty-eight miles of which were unsurfaced. For mining and processing rutile you need mills, a power station and ancillary buildings, an earth dam, roads, houses, sewage works, water mains – and a cross-section of skilled trades to provide them. 'By February 1966, we had a staff of fifty on site, plus seven or eight wives and a number of children.'

How to accommodate them? bring air-conditioned caravans from U.K. with single men messing together? No: there was a company not too far away which manufactured prefabricated wooden bungalows. Central messing, or small messes for single men; what is the ideal number for sharing a house? Civil production foremen, engineers, and office staff seem to prefer living in small groups; steel erectors, welders, and electricians generally vote for a communal mess. As for accommodation, 'four men living together appear to be a reasonable compromise between cost and comfort'.

Houses were built of local hardwood, impervious to termites, two or three feet off the ground on concrete blocks: this makes drainage easier in the rainy season, and 'keeps out most crawling creatures'. Windows were mosquito-screened. Married quarters had electric water heaters for baths and laundry, and refrigerators. The project manager's house had an extra bathroom and bedroom for 'visiting firemen'.

A club house is essential, with a small stoep outside for cinema shows (rather old films, but none the worse for that), a games room, an area for dancing, and a paperback library. Among other needs are a dispensary for the site nurse, a food shop, 'a bigger septic tank than you think you'll need', a telephone exchange, and short-wave radio sets for contacting base. With a bit of luck, you can sell the lot to your client when you leave.

Much more primitive was the 'hole in the ground' Frank

Taylor found, many years ago, when an oil pipeline was being driven through the southern province of what is now Tanzania:

'one of our men had established a grass hut, which he built himself with materials valued at 25s., and in this he had attained a degree of comfort which had to be seen to be believed. Although his nearest white neighbour was sixty miles away, he was able to live happily off the country – shooting guinea-fowl, collecting fruit. When I arrived at about 7 o'clock one evening he was out in his little grass shower with his boy pouring buckets of water over him from the open top to wash off the red laterite dust which had caked him during his day's work of driving traces through the jungle. These are the men who are the backbone of the British construction industry.'

II

A Home of One's Own

WHETHER you believe in astrology or not, it is interesting to consider Frank Taylor's birthday – January 7 1905. He is what a student of horoscopes would call a 'Sun Capricorn'. Capricorn is an earth sign; and it cannot be denied that Frank Taylor has shifted many millions of tons of earth in his time. When the astrologer goes on to say that such a man shows 'a practical application to the concrete', he or she probably does not mean the kind of concrete *we* mean; but when we are told that the Capricornian 'provides shelter for others, striving to achieve security by activities in the outer world'; has 'responsibility and a sense of timing', and 'wins through against all obstacles', we begin to recognize someone we know.

The village of Hadfield, Derbyshire, is eight miles from the Peak, two and a half from Glossop, and two from the Devil's Elbow. It is high enough in the foothills of the Pennines to be severely cold in winter. Here Francis Taylor and his wife Sarah Ann (*née* Earnshaw) kept a small fruiterer's shop in the front room of No. 33 Station Road, a five-roomed terrace house. Their son Frank helped in the shop almost as soon as he could walk, and eventually went to Hadfield Castle School (now Castle Secondary Modern) – a 'Mixed Council School' where he showed average aptitude, and perhaps above-average leadership. In 1959 he had the satisfaction of going back there, as a local boy who had made extremely good, and distributing the prizes on Speech Day.

In his address to the school, he said he had had the good fortune to be born 'of wonderful parents in a fine Christian home'. He remembered how his parents had worked a six and a half day week from 5 a.m. till 10 or 11 at night. 'Even on Sunday morning, when the shop was closed, people would come round to the back door for potatoes or strawberries they had forgotten on Saturday. The kitchen had a roof which leaked like a sieve, and there was a cast-iron boiler by the side of the fire which one had to fill with a jug to get hot water.'

In winter, he often went to school by sliding down Wesley Street in the snow. There was neither bus nor bicycle to take him there: 'I think walking to school plays a great part in making a healthy child – today, I believe, some children even take taxis!'

The headmaster, William Whiteley, known as 'Ben', was a remarkable man whose long-term influence is felt in Taylor Woodrow sixty years later. 'He was very much ahead of his time,' Frank Taylor says. 'He knew he only had us until we were thirteen, and that none of us would be going on to higher education. It was no use cramming us with facts, and he didn't make us learn things by heart as most schools did then. Instead, he taught us something much better: *how* to learn, how to help ourselves, how to acquire knowledge, and where to find it. He was also very keen that we should *speak* well and write our own language fluently.'

Frank also learnt to swim, but not at the school – State schools in those days had no such amenities as swimming pools. The boys were allowed to leave early in the afternoon in order to visit Glossop baths. It was a rule that the boy in charge of the party should be able to swim, since he was responsible for the others. Frank volunteered for the job.

'But you can't swim,' he was told.

'No, sir, but I'll learn this evening!'

Being Frank Taylor, of course he did.

It was a singing school, and Frank enjoyed being in the school choir, which had the honour, every year, of taking part in the Buxton Music Festival.

Frank's father, a Lloyd George Liberal and a man of strong principles, was not a great church-goer: it was his mother who saw that young Frank went to Methodist Chapel and Sunday School.

Two incidents of his boyhood have become part of the Taylor saga. 'When I was nine, my father caught me smoking and strapped me. I've never smoked since.' Perhaps it was not only the strapping: he simply didn't like smoking, though he is extremely generous in pressing cigars upon visitors.

The second incident happened a year later. 'A friend and I were very fond of ginger beer, which used to be kept in big stone jars in his mother's house. What we didn't know was that some of the jars were being used to store home-made dandelion wine. Have you ever tasted that stuff? It's pretty potent! We'd drunk about a jar each before we realized something was wrong. We both had to be carried to bed.'

Since then, Frank Taylor has never touched alcohol. Was the dandelion wine the only reason? Pressed, he admits: 'Well, no – I had a relative who became an alcoholic. That's a scarring experience when you're very young.'

When Frank was ten, something happened that was to change the family's life. A baby brother was born, but died eight months later. His mother was very upset, and in poor health – she suffered from bronchitis, and needed a milder climate than the Derbyshire hills could provide. So his father took her to Blackpool for a holiday, leaving young Frank in sole charge of the shop – at the age of eleven.

This did not mean merely serving behind the counter. It meant getting up at 4.30 a.m. to drive the horse and cart to

Manchester market, eleven miles away, collecting the fruit
from the wholesaler, and getting back to Hadfield in time to
open the shop at eight.

Mr and Mrs Taylor returned from their holiday determined to move to Blackpool; but this was not to happen for
another three years, during which young Frank became the
complete 'half-timer' – going to school, but also looking after
shop, horse and cart, and often doing a fruit round as well.

At thirteen he left school and, typically, sent himself to
night school: Ben Whiteley had shown him that education
must go on if you are to equip yourself for better things.

At last the great plan was achieved; the Taylors moved to
Blackpool, where they took a rented house in Grasmere
Road, and set up a new fruit business – a wholesale one this
time.

Frank, at fourteen, took his first job outside the family
business: he joined Edward Gregson, another wholesale
fruiterer in the district, as general assistant. He also
bought a motor bike with £20 he had saved up (he was
always a saver). To ride a motor bike he had to have a
licence – on which the words 'and/or a motor car' had, by
some oversight of an overworked post office official, not been
deleted.

It was strictly illegal, of course, to drive a car at the age of
fourteen with only a motor cycle licence, but this didn't
bother young Frank. He had by now made friends with
Andrew Isherwood, Mr Gregson's lorry-driver, who taught
him to drive a Model-T Ford which was used to fetch
produce from Gregson's farm. Those who have driven a
Model-T will know that you had to change gear like a motor
bike with a foot-lever, down-one up-three.

Mr Gregson was furious when he found out. However, he
too was an opportunist, and a day or two later told Frank
gruffly: 'Well, now that you can drive, you can take the

lorry to Manchester.' This was Frank's first experience of
double-declutching and a lever gear-shift, and it was a seven-
ton Pagefield lorry. Somehow he brought it back in one
piece. 'The first ten miles were the worst,' he says philosophic-
ally.

Cars have always fascinated Frank Taylor: around Hadfield
he had often seen the chassis of new cars being tested, and his
ambition at this time was to be a racing motorist.

Blackpool in 1921 was not so very different from today.
The 520-feet Tower, a portent of some of the high buildings
Frank Taylor was destined to erect, was there; but not for
another seven years would Reginald Dixon take over the
mighty organ at the Ballroom. Illuminations, Wakes weeks,
fresh air and fun, Albert and the Lion – millions of holiday-
makers invaded the town in summer; but in the winter
Blackpool people, as at all seaside resorts, almost literally took
in each other's washing.

In Blackpool, as everywhere else in the country, there was
a great hunger for houses. There were millions of unemployed
who could have been put to building them, but the planning
and the money were not there. 'A home of one's own' was
the dream of every ex-soldier. People even sang songs about
it – 'just a love nest, cosy and small', and the Co-optimists'
'In our little garden subbub, away from the noise and
hubbub'.

At Welwyn Garden City, houses designed by architects
like Louis de Soissons, with loggias, roof-gardens, and
pergolas, were going up in the great new housing experiment
of the day; but since ordinary people could not afford them,
they were no solution to the nation's housing problem. Nor
were Bush House, India House, and Australia House,
Aldwych, that great tribute to the Commonwealth, then
London's most spectacular piece of urban renewal, for whose
foundations John Mowlem was excavating.

Contemporary issues of *The Builder* had leading articles on 'The Housing Chaos', and blamed the Government. Sir Alfred Mond, Minister of Health, had fixed a limit to the Government's housing schemes. Lloyd George, said *Punch* bitterly, 'has resolved to find a land fit for heroes to emigrate to'. Concrete was much in the news, but it was to be used for roads 'now that there are so many vehicles'. The very latest equipment was the Tonkin concrete mixer ('works by hand or motor') which extruded the stuff on a sort of Archimedes-worm principle. It cost £100.

To be fair to the Government, it did have other things to think about. The troubles in Ireland were at their worst. There was a coal strike which not only crippled industry but brought the five-year record run of *Chu Chin Chow* to an end at last. The railways were still not back to full service after the War. There were still army huts in the Horse Guards waiting to be dismantled.

Keynes had just published *The Economic Consequences of Peace*, and Bertrand Russell *The Theory and Practice of Bolshevism*. President Harding had just repudiated the League of Nations in his first message to Congress. Britain's pride and joy was the *Aquitania*, which 'is so wide that she could not be squeezed into Northumberland Avenue'.

Colonel Fawcett was in Brazil, but not yet lost. There was talk of an Everest expedition. *Punch* made jokes about plus-fours, bobbed hair, psycho-analysis, Mr Bottomley, plumbers forgetting their tools, and war-profiteers falling off their horses in the hunting field; and there were cartoons about young men with names like Lord Algernon Montmorency saying things like 'This is a toppin' fox-trot, isn't it?' (the tune was probably *Japanese Sandman*).

One could escape into cricket, with Hobbs, Hendren, Strudwick, and Woolley. Or into the silent cinema, where women were swooning before Rudolph Valentino in *The*

Sheik. Or into the West End theatre, where Maugham's *Circle* had just been produced, Noel Coward (at twenty-one 'one of our precocious popular favourites') was playing in *Polly*, Gerald du Maurier in *Bulldog Drummond*, and Isobel Elsom in *Up in Mabel's Room*. Owen Nares, the only other matinee idol who could compete with du Maurier, had just been fined £1 for driving a car at twenty-nine m.p.h. However, predicted *The Times*, 'next year the speed limit for private motor cars will be abolished'.

In Blackpool, the Taylors were tired of living in a rented house. Houses were difficult to buy: not only was the housing shortage worse than anything we can imagine today, but prices were high and 'easy terms' almost unheard of.

'Why,' asked young Frank, 'don't we build our own?' Father was doubtful, but he had already learnt that Frank's judgment was worth backing. Frank had £30 in the bank, and was prepared to risk the lot. Father was willing to put up £70. It was then that Frank, aged sixteen, performed the first of many miracles with banks. He persuaded the manager of the District Bank to lend him £400. He has always pretended to be 'the world's worst salesman next to Frank Woolworth', but there must have been something quite abnormally magnetic about him to be able to pull that one off. At the time, he knew nothing about building. 'Being naïve,' Frank Taylor says today, 'I used the loan to buy materials instead of land! I didn't know any better.'

A youth named Turner, a friend whom he had met at Sunday School, taught him how to mix concrete and mortar. (Leslie Turner, nearly sixteen years later, became Taylor Woodrow's No. 1 man in America.) One of the experiences that enables Frank Taylor to understand the building trade from the tea-boy upwards is that he has personally done every job – carpentry, bricklaying, hod-carrying, the lot. He knows what tea made in a cement-bucket tastes like.

He was in fact building *two* houses, not one – a good, sound, semi-detached pair that can still be seen in Central Drive, Blackpool. One was to be for his parents, the other for his uncle, Jack Woodrow. 'I got the plans approved myself, I set them out myself, and I did all the jobs – anything that was needed in order to get those houses built as quickly and economically as possible. Not only did I learn the essentials on the technical side, but in working with people on the actual buildings I got to know them, to understand them.'

Long before the roof was on, passers-by were making offers for the houses. Frank eventually succumbed to temptation, and sold them for £1,000 each – an enormous price for those days. It was not clear profit, of course; but it was a 100 per cent profit, since land, labour, materials, and repaid loan had cost him altogether £500 each; and it decided Frank that building was the career for him.

Father again was dubious; perhaps a little disappointed. He had set his heart on sending Frank to California to learn the fruit business on a grand scale, to become a big grower, importer or canner instead of a small wholesaler. But 'My father never forbade me to do anything enterprising,' Frank Taylor remembers.

The bank manager was doubtful, too, and advised him to take his profit and get back to the fruit business. Frank's reply was to ask for another loan. Two houses soon became twenty.

When he was eighteen, his solicitor Luke Matley said to him one day: 'My son Peter will be twenty-one in three weeks. We're giving a party for him – you'll come along, won't you?'

'I'd be delighted to,' Frank said. 'After all, I'll be twenty-one myself in three years' time!'

Luke's jaw dropped. 'You'll be *what*?'

'Twenty-one. I'm eighteen now.'

'You mean – I've been aiding and abetting a minor to buy

and convey land under age?' Luke was the picture of an out-
raged family solicitor. 'It's illegal! This simply won't do, you
know. You must take an older man into partnership, otherwise we'll both be in trouble.'

'What about Uncle Jack?'

'He'll do.'

So Jack Woodrow (who died in 1929) lent his name and a
private company was set up, with a tiny office on a housing
site in Blackpool. 'Taylor & Woodrow': for some years
Frank was uncertain whether it was a good name. 'People
mixed us up with Woodrow Wilson and Taylor Walker's
beer. It's too late to change it now, though!'

Another loan from the bank – this time to buy land and
build eighty-five houses. What kind of houses were they?
Frank Taylor is scornful of the jerry-building which was rife
in the 1920s; it was, he thinks, not only immoral but stupid;
but there was less of it than people think. A house is the
biggest purchase most of us make in our lives: 'If you don't
build houses with good materials, so that they are warm and
comfortable, people won't buy them.' In the north-west he
was building the kind of houses people wanted in those days –
with small back and front gardens, three bedrooms, a bathroom, a hall, a drawing-room, a dining-room, and a kitchen
with typical north-country outhouses for coal and washing.
(Home laundry in the 1920s was done in a 'copper' with a
curious wooden tripod for pounding the wet linen known as
a 'dolly'.)

Frank Taylor's first employee was a carpenter named
Harry Smith, who afterwards went into horticulture and lost
touch. One of the very earliest team members on the pay-roll
was William (Billy) Bee who has been with Taylor Woodrow longer than almost anyone else – since 1928 – and is
today a director of Taylor Woodrow Homes Ltd.

Another was Tommy Fairclough, who when he was

twenty was recommended to Frank Taylor as 'the best carpenter in Blackpool'. He was destined to become managing director of Taylor Woodrow Homes. Known as 'Tommy Builder' to distinguish him from another Tommy Fairclough (a drainage expert, also from Blackpool but unrelated, who was called 'Tommy Drainer'), he died in 1959. His son, yet another Tommy, is now sales manager of Taylor Woodrow Homes. 'Tommy Drainer' is now retired, and lives in Berkshire. 'Tommy Builder' had specialized in greenhouses until he got his first house-building work with Taylor Woodrow 'at £11 a house, including fences'.

He teamed up with men like Andy McClelland and Jimmy Banks, who laid bricks and plastered on labour only; and Vic Roberts, the decorator (now with Myton). 'Mr Taylor himself,' Tommy used to reminisce, 'was always there from early morning until late at night, unless he had to push off somewhere for short periods in his bull-nosed Morris Cowley.'

Sometimes they were short of labour: then 'Mr Taylor would think nothing of standing all day over a lime box slaking lime, or taking up hods to the bricklayers.' Once Frank Taylor knocked Tommy up at night because there was a danger of the newly erected walls of a four-block being blown down by a gale if they didn't get the roof on quickly. 'As was usual with all of us,' Tommy said, 'I just looked at him and said "yes, Frank", and away I went, taking a couple of blokes and a cart lamp, and throughout the night we worked in a sixty-mile-an-hour gale; and when the job started at 8 o'clock next morning, the roof was all ready for the tilers to start work.'

Everyone looked forward to Friday night: 'Mr Taylor would drive up from the bank in the Morris Cowley with his little attaché case, and pay every sub-contractor in turn.'

For a successful north country builder to move south in

5(a). Expansion into Africa. Takoradi Harbour extensions, Ghana.

5(b). Bowater House, Knightsbridge.

6(a). Central Terminal buildings and main access tunnel, Heathrow. The start of twenty years of work at the airport.

6(b). The Plant Efficiency Cup for 1953 was awarded to the Heathrow Airport construction team. Left to right. Tom Freakley, Cyril Bayton, Harry Hooper, Reg Stinchcombe, Bob Hammond, Sandy Cheyne, and Tom Reeves.

7(a). 1955. Return to Blackpool. Tommy Fairclough looks on as singer Joan Regan opens the demonstration home at Thornton Cleveleys.

7(b). An interesting project in South East Asia. The new University at Rangoon.

8. 1956. A milestone in history. Power flows into the National Grid from the world's first nuclear power station as H.M. Queen Elizabeth II inaugurates Calder Hall.

the 'twenties was almost unthinkable. What would become of him among all those smooth sharks of the south? It was an opportunity, a typically quick decision, a hunch, and a calculated risk that changed Frank Taylor's course in 1930.

Nat Fletcher, director of Taylor Woodrow Services Ltd. and head of publicity for the group, remembers his first meeting with Frank Taylor at about this time. Nat, then a boy of ten, was spending his holiday at Blackpool, staying with his uncle and aunt. The doorbell rang and Nat answered it. He found himself staring into the 'blazing blue eyes' of a sandy-haired young man of twenty-five who seemed very excited about something. Nat's uncle, an estate agent, had done a good deal of work for Frank Taylor's various housing estates. The two men were closeted together in earnest conversation for a long time.

After Frank Taylor had gone, Nat heard his uncle say to his aunt: 'Well, Gertie, it looks as if we'll be going to London.' He had been mesmerized into taking the same risk: he sold his Blackpool business and came south to throw in his lot with Taylor Woodrow. So did Tommy Fairclough and several others – Tommy on his New Hudson motor bike, on St Valentine's Day 1930. Once again he had said 'Yes, Frank'.

What was this new scheme in the south? Frank Taylor knew an engine fitter in A.E.C., a firm making buses and lorries in the London fringe. He had seen Taylor Woodrow's advertisements while on holiday in Blackpool. He had probably also seen advertisements which had been carefully placed in the south – 'Taylor Woodrow Estates – Best in the North-West!' One day he wrote to Frank Taylor out of the blue and said: 'Look, the whole factory's moving to Southall – it's the other side of London, in Middlesex. We'll be needing hundreds of houses for our employees like the ones you've been building in the North-West. How about it?'

c

Frank Taylor took the next train south and had a look at
the land around Southall. The nearest to the factory site was a
curious bit of ground at Grange Park, in Hayes End. Grange
Park had 120 acres, and it was not hard to see why it was no
longer viable as farmland. To any conventional land-agent,
it was no good. It sloped the wrong way. It was undrainable.
The main reason why it was available was that other developers had turned it down.

'No Dutch engineer would have hesitated to buy it,' Frank
Taylor recalls. 'You can't waste land just because it slopes the
wrong way. Sewage can be pumped *up* into the main drains,
can't it? I reckoned that if I put in a small pumping station to
serve the estate it would work out at only thirty shillings per
house.'

So he marched round to the local branch of the Midland
Bank and set about trying to persuade the manager that they
were both on to a good thing. For a man who constantly
belittles his own ability as a salesman, he pulled off another
miracle. He came out of the manager's office with overdraft
facilities of £15,000 on security of the land, and up to
£25,000 on top of that, based on a rate of £300 for each
completed house. That bank manager was Billy Burns, now
retired and living in Church Road, Hayes; and that branch of
the Midland still holds the Taylor Woodrow account.

'Of course it was a tremendous gamble.' Frank Taylor now
says. 'Everyone in the trade thought I was crazy. Taylor
Woodrow must go under, they said. If the simplest thing had
gone wrong, or if the bank had got cold feet, I would have
gone bust.'

Now began what the late Sir Harold Bellman, chairman of
Abbey National and afterwards a director of Taylor Woodrow, who knew him in these early days, called 'F.T.'s blood-
less revolution – the £450 house'. Some of the houses, it is
true, were £550 freehold – but you could get one for £25

down and repayments at 14s. 3d. a week for twenty-three years.

As always, Frank Taylor did his publicity homework before building a lot of houses. 'I opened a showhouse, and before the week was over I had sold fifty houses.' He had also anticipated the age of mass motoring; each Grange Park house had space for a garage, and the garage itself cost £30 extra.

The 1,200 houses on the estate took three and a half years to build; and meanwhile Taylor Woodrow were working on other estates in the Home Counties too.

Luckily, we have an eye-witness account of the Grange Park Site as it was in 1931. Once again, it is from the late Tommy Fairclough, who in 1951 contributed a series of reminiscences to *Taywood News*, the Taylor Woodrow house magazine.

Hayes End 'was chiefly open fields, and farmhouses'; and 'trams were the main highroad's chief transportation'. Work began on the footings of the first four-block in what is now known as Lansbury Drive. Tommy fixed the first floor joists – 'only a day's work for two of us' – and was told by Frank Taylor to go and enjoy himself in London until the brick-layers had the houses ready for roofing.

'A joiner's shop had been rigged up in the farm barn for making all the prepared joinery, windows, staircases, internal door frames, etc. One of the joiners was a young man named Reg Wreford' (now a director of Swiftplan Ltd.).

Another farm building had been adapted as a store for ironmongery and timber, and a petrol-driven circular saw stood by. A temporary office had been set up in the farm-house.

A firm of estate agents had arrived to sell the houses. 'Small SOLD boards were visible on hundreds of plots, which was not surprising when the selling prices of the houses were from £445 freehold, which included all road charges and legal fees.'

Lansbury Drive was well under way; so were Woodstock Gardens and Woodrow Avenue. Several other bricklaying gangs were at work on these, led by Billy Bee (whom we last met in Blackpool).

'Billy Bee was the only one in charge of any of the brick-work gangs who could brick out the houses at the price given, which was £108 for a four-block . . . it was not unusual for him in those days to lay as many as 2,000 bricks a day, and if any of his bricklayers could not produce at least 1,000 a day, they didn't stay long in his employ. It was quite common for him to start a bricklayer at 8 o'clock in the morning and for the man to be seen walking down the road at ten minutes past.'

By April 1931, houses were being completed and passed for habitation at the rate of a four-block a day. Among many of the original Blackpool gang were Bob Hirst, a bricklayer who became a director of Taylor Woodrow Construction Ltd. and Andy McClelland who had risen from sub-contractor bricklayer to general manager of the project. Frank Taylor himself was on site all day, arriving in his latest car, described as a 'fabric-bodied Marquet'. In the evening he would often fill his car with the team members and take them to Lane's club, Baker Street, for the all-in wrestling or to the Royal Albert Hall for the boxing.

The following year Frank Taylor changed the Marquet for a Morris Isis sports saloon. He now had a chauffeur named Norman Janney, in whom other abilities were quickly dis-covered: he became a plant supervisor.

Billy Bee, now a general foreman, had moved on from Grange Park to complete his first estate – 136 houses, includ-ing roads and sewers, at Bilton Road, Perivale, and in the next few years was to build estates at Kenton, Stanmore, and Edgware.

It is now, in 1932, that we first discover the name of Leslie Olorenshaw, who had joined the company as a wages clerk in the office of John Hanson, then assistant company secretary. John Hanson went on to become a founder-member of the parent board when the company 'went public' in 1935, and was eventually managing director of Taylor Woodrow Construction and chairman of Myton before his retirement in 1962. 'L.O.' was eighteen when he joined and he earned £3 10s. a week – good pay for those days – though it was not long before he had risen to £10 a week. Leslie Olorenshaw is today joint deputy chairman of the parent company, chairman of Taylor Woodrow Homes, and director of Taylor Woodrow International. He is responsible for all group activities overseas.

Next year, the Corinth cinema was built at Hayes – in four and a half months. By 1934, yet more estates were opened up in Middlesex, Surrey, and Wiltshire, some of them administered by subsidiary companies with names like County Homesteads and Sunshine Homes, which were eventually amalgamated into Taylor Woodrow Estates.

Two long-service team members joined in this year – Frank Pursglove, as Frank Taylor's private secretary (he is now in charge of all group insurance), and Doris Goldsby, who came to take over the company's rapidly extending telephone system. Known throughout Taylor Woodrow as 'the girl with the golden voice', she is now Mrs Sullivan, in charge of the main reception switchboard at the Park Street headquarters of the parent company.

In the 1920s, young Frank Taylor had said to himself: 'I'll make £10,000, and then I'll retire.' He had had a vague idea of taking his capital to somewhere like South Africa and growing peaches. His fruiterer father would have approved. So, surely, would his Blackpool bank manager, who had always regarded building as a gamble.

It was not ambition alone that pushed Frank Taylor towards the first £10,000 – or indeed the first £100,000. 'I realized that it wasn't I who had made that money,' he says today, 'it was a team of people. The people around me. They had started to look to me as the entrepreneur whose job was to create business, to keep it coming in, to see that it was profitable, to keep the team going. So from then onwards I was hooked – and was glad to be hooked with this team of people. The only difference today is that the team has grown very much bigger.'

III

Undertones of War

TAYLOR WOODROW, in 1935, had reached the stage of its development when it was ready to come to the market. So the team became a public company with a capital of £3 million, and a board consisting of Arthur Collins, Frank Taylor, John Fenton, Bert Rigg, John Hanson, and A. E. Aldridge.

There were a number of good business reasons for this, among them Frank Taylor's restless eye which at the moment was fixed on America. He also wanted public works contracts, which would need special financing.

America, in the 1930s, needed houses as badly as we did, but ordinary people, even less than in Britain, could not buy them because, under the law as it stood then, they had to put down one third of the price as deposit; whereas in Britain, as we have seen, you only had to scrape together £25 and the house was yours for 'easy payments'. True, America had savings banks, rather like our building societies; but a 'home of one's own', especially in a country where so many people lived in apartments, was still difficult.

What Taylor Woodrow were waiting for, in fact, was The Federal Housing Administration, a Government financial insurance scheme which allowed loans up to ninety per cent. This came in 1936. At last America was ready for Britain's accumulated experience of building housing estates – British know-how (Taylor Woodrow's in particular) of construction applied to American architects' knowledge of local conditions and materials. Frank Taylor lost no time in getting there.

But before we follow his fortunes in the United States, which he first visited in 1936, we must go back a little way to see how the company was faring at home.

New offices at Southall had just been finished and staff had moved in from Uxbridge Road, Hayes (next door to the Corinth cinema) to Adrienne Avenue.

These new premises were grouped around a stretch of the Grand Union Canal, known as Engineers' Wharf (after the men who built the canal), and today the Taylor Woodrow building here actually straddles the 'cut'. By pure coincidence, there was a little old beer-house adjacent, lit by oil lamps, called The Civil Engineer. When later they needed the land it stood on, Taylor Woodrow bought it, and as part of the agreement built for the brewery a brand new pub a little farther away – but still conveniently near for Southall staff.

Among those who joined the expanding staff at this time were an invoice clerk named Leslie Moere, afterwards managing director of Builders Supply Company and now chairman of Swiftplan Ltd; another invoice clerk named Peter Boorer, later chief buyer for Taylor Woodrow Construction and now a director of Taylor Woodrow Services; and a wages clerk named Ronald Copleston, now a director of Taylor Woodrow Construction and managing director of Taylor Woodrow Services.

The summer of 1935 is remembered by old hands at Taylor Woodrow as a season in which the company broke house-building records. In June they were working on Cranford Park Estate, Hayes, when instructions came from Frank Taylor to complete a pair of semi-detached houses in the least possible time. How quickly, the experiment was to discover, *can* you build a house? At that time, the record speed of ten days was held by a competitor. Tom Fairclough, in charge of the project, carried out a piece of organization which resembles what is nowadays called 'critical path scheduling'. A programme was devised, all materials and labour lined up. 'From the time the first sod was cut on site, to its being certified fit for habitation by the Council, was exactly eight days.'

The brickwork sub-contractor, Bob Hirst, had the pots on the chimneys in two and a half days. Vic Roberts, in charge of decorating, had to do his priming by candlelight in order to achieve his time-target.

More and more estates were built, and in the summer of 1936 the rate was over fifty houses a week – at Melksham, Trowbridge, Chippenham, and Calne (all in Wiltshire); at Edgware, Ruislip, Eastcote, and Sudbury, and as far east as Rochester, and as far west as St Budeaux, Plymouth.

A checker named Don Fidler (today he is chief cashier at Southall) joined the firm, and an accountant named Reg Heasman (now director of the parent company and chairman of Myton and of Taylor Woodrow Services). Reg Heasman was not unknown to Frank Taylor, having been Secretary (at the age of fourteen) of the Methodist Sunday School they had both attended in Blackpool seventeen years before. Reg Heasman's first job with Taylor Woodrow was as secretary of I.W. Properties, a subsidiary which re-let the houses of building society defaulters.

Despite all this activity, it was plain that the housing boom in Britain was losing speed. 'We had a lot to offer America in the way of housebuilding,' Frank Taylor says. 'We were ahead of them in our long experience of design and specification, laying out housing estates, and buying land.'

Almost the first thing Frank Taylor did on arriving in the States was to buy a golf course on Long Island for one million dollars. You can build an enormous number of houses on a golf course. 'I wish I had held on to the land,' he said many years later. 'It would have been worth fifty million dollars now.'

There were several things the Americans liked about what they thought of as 'the English style' in building: Tudor-type half-timbering, and roads with traditional English names such as Gloucester and Kenilworth. You can still see this sort of thing in the suburbs of Los Angeles, where *cul-de-sacs* full of 'little boxes' tend to be given names like Chevy Chase and Northumberland Avenue. 'We were also the first estate builders to introduce curvilinear streets,' Frank Taylor recalls. 'I wasn't too keen on them, but they were very popular.'

Taylor Woodrow were deliberately cutting across the American 'gridiron' layout, which had been traditional in real estate, probably because it was thought to economize land. They were giving American small-home owners a new sense of privacy, a garden city touch. Straight streets were death-traps, as the number of private cars increased, so that curved side roads and *cul-de-sacs* could also be said to make life safer.

There were however problems on which Taylor Woodrow needed local advice. They were now, with their first overseas subsidiary company, working in a very difficult and sometimes extreme climate, and this affected the materials used: for example, it was often safer and cheaper to use felt roofing instead of tiles.

Building on the golf course estate went on for the next three years, right up to the outbreak of the Second World War. This was the foundation of much bigger interests in America, and of associated companies such as Greenpark Construction, Greenpark Essex, and Greenpark Sussex (still the English touch, you observe).

Two important personalities in Taylor Woodrow's American debut were Jack Kony and Tom Keogh. On this side of the Atlantic Jack would be called an estate agent, the man who actually sells the houses. Over there, he was generally called 'just the broker', being thoroughly familiar with sites, values, public utilities, building codes and costs labour, financing, and legal conditions. Tom has been the company's attorney in America since the beginning. To their wisdom and experience Taylor Woodrow owe a considerable debt.

Back in England, the company seemed to be having slight troubles with bricklayers. At all events, the then house journal *T.W. Chronicle* (April 1936) had a leading article every month addressed to 'Bill, the Bricklayer', telling him what a good chap he was, and how silly it was to be 'suspicious of the Bosses, who are not too bad when you get to know them'. There was of course a message from Frank Taylor about the importance of team-work; a reminder of how cushy modern brickies have it compared with the poor devils who humped stones for the Pyramids and Stonehenge; lots of sport; and – inexplicably – a very serious article on 'Courtship in the animal world'.

Incidentally, this was not the first house-journal; a previous one, called *Action*, had had a brief life in 1935, and is of interest today for an astonishing coincidence – or possibly a consistency of thought. The August issue contained an article, 'It's up to You', illustrated by *three* men pulling on a rope. Not for another twenty years was anyone to think of the 'Four men' symbol of team-work. Yet here it was in 1935, summed up in typical Frank Taylor style with the exhortation: 'Pull together. If you can't pull, push – and if you can't push, then get out of the way.'

We have seen that future expansion did not lie in housing alone. Taylor Woodrow now decided to go into the field of contracting and civil engineering, and in 1937 Taylor Woodrow Construction Ltd. was formed with John Hanson as managing director. Tom Fairclough became managing director of the company carrying out the development of the housing estates. The estates were now spreading back from south to north, and so this division of the company had its headquarters near Birmingham.

The construction company now looked around for something to construct; and its first contract was to build a public convenience at Harrow Weald. It was by no means easy to get. 'I had the devil's own job to convince Mr Rockham, the borough engineer, that I was a fit and proper person to build it,' Frank Taylor recalls. 'He kept asking "But have you ever built one before?"

'Years later, it was much easier to get a contract worth millions to build the first nuclear power station, because nobody could ask us whether we'd ever built one before. Experience is all very well, but you've got to begin somewhere.'

Any pilgrim who may wish to pay tribute to this early example of Taylor Woodrow art will be glad to know that the comfort station is still there and in excellent working order. It cost £1,500.

There was not an immediate rush of work for Taylor Woodrow Construction, and housing estates were still the mainstay of the business. In 1937 and 1938, new ones were begun at Acocks Green and Perry Barr, Birmingham; Walsall, Scunthorpe, Plymouth. But the Construction Company had pulled off a large contract at Rosehill, Oxford; and it had acquired a general manager in the person of A. J. Hill, now joint deputy chairman of the parent board, and chairman of the Construction Company. 'A.J.', an ex-Sir Robert McAlpine man, was a young engineer whose experience of heavy civil engineering was to prove invaluable to the group. Among other things, he had worked on the naval base at Singapore, which was where he had met a young accountant named Ron Copleston.

The Second World War was very near, and around Munich time Taylor Woodrow Construction became very active indeed building militia camps – one at Colchester, under John Hanson's direction, and another at Denbury, near Newton Abbot, with A. J. Hill in charge. Some of this early work was shared with Wimpey, with whom Taylor Woodrow went fifty-fifty in a company called Western Engineering, formed for the sake of speed.

'We built them at lowest cost and quickest,' Frank Taylor recalls, 'for Brigadier Bliss at the War Office. He didn't seem to know *how* quick we were. I saw him just before Colchester was due to be completed. 'When do you think it will be ready?' he asked. I told him it had been occupied for a week!

On September 3 1939, within an hour after Neville Chamberlain's declaration of war, Frank Taylor telephoned instructions throughout his organization that all houses that were anywhere near the roofing stage were to be completed as quickly as possible, and that no further work was to be done on the remaining properties. The estates in America could be finished, but these too had to be hurried up.

A full board meeting was held on September 6, and the company's plans 'for the duration' were outlined. Among them was the evacuation of the accounts, quantity surveying, and insurance departments from Southall to a 500-house estate which Taylor Woodrow had recently built at Kidlington, Oxon.

This happened in 1940, under the joint direction of Reg Heasman and Ron Copleston.

At headquarters, Taylor Woodrow's own *Dad's Army* was organized, fifty strong, with A. J. Hill as commanding officer, and Frank Taylor himself (who had taken a small house in Woodstock Avenue, Southall) as chief fire-watcher; and at Kidlington, C.S.M. Copleston was in charge of the local home guard from his staff of forty, assisted by Mr Dupont, landlord of the Dog.

Frank Taylor, now thirty-four, volunteered for military service, but was told that he and some of his team of similar age were wanted for war work. Many Taylor Woodrow people joined the Forces, including Les Olorenshaw, who soon found himself commanding a tank in the 8th Army in Egypt, Libya, and Palestine. While he was away, Reg Heasman took over as company secretary until 1942, when he handed over to Lewis Daniels.

The war was to raise Taylor Woodrow from a prosperous building firm specializing in housing estates to a place among the giant names in the contracting and civil engineering world.

Even so, contracts did not come easily at first. In the two years before war broke out, Frank Taylor had sometimes wondered whether there was something wrong with his business-getting methods.

'Somehow, we couldn't get anything out of the Air Ministry,' he says. 'Our first tender was for an installation outside Chester. Our price was £1,750 – £409 below cost. We were turned down. We tendered for another job at £2,100 – and *that* was £400 below cost too! Again we were turned down. I couldn't understand why.

'So I got Sir Reginald Clarry, then M.P. for Swansea, to go to the Ministry and try to find out why we weren't getting any work from them. Their answer shook me: "Taylor Woodrow," they said, "aren't competitive!"

'Imagine how much our competitors must have been undercutting! Still, the next time we tendered it worked, and after that we were on the "unlimited list".'

Frank Taylor has another story about civil servants, similar to the episode of the Harrow public convenience.

'We approached the Ministry of Works to ask if we might

tender for an ordnance factory. "Oh, but," said the civil servants, "you've no experience of building ordnance factories!" Well, eventually we got that contract too.'

There were also gunnery camps in Cornwall – Bude, Newquay, St Agnes – all built with remarkable speed; and anti-torpedo boat gun emplacements – nests of buildings which included power generators – at Portsmouth, Gosport, Hurst Castle, Freshwater, Plymouth, Falmouth.

Taylor Woodrow were soon building more than camps for the Army, and much of their work was top secret. They were constructing land and sea defence works along the East Coast for the C.R.E. at the War Office.

'One of our tasks,' Frank Taylor remembers, 'was to build a Bren gun post. Somehow, the drawings looked wrong. I asked for an actual Bren gun so that we could make sure the pill-box was the right size and shape. "Sorry, old boy," the military replied, "we haven't got one!" So we knocked together a model Bren in wood.

'What nobody had remembered was that the magazine of a Bren gun sticks *up*, not *down* – so we nearly built a post from which the gunner couldn't see his own field of fire!'

One of the difficulties of those days was that all signposts, indeed everything legible, had been taken down for security reasons, and you never knew where you were or how to find your destination. 'There wasn't even a sign that said *Gentlemen*.'

The construction industry at this time was very dependent on labour from the Republic of Ireland, which was neutral. Frank Taylor, needing a thousand men, sent an Irishman named Dwyer to recruit them. They arrived in good order, complete with their own Roman Catholic priest.

'The rate for the job was thought to be 3s. 7d. per hour,' Frank Taylor recalls, 'and this was what other construction companies were offering; but T. W. recruited on the basis of

3s. 5d. per hour to make sure. When the other companies' men got to England and found that the official rate was 3s. 6d. they were angry and discontented. T.W., who had offered 3s. 5d. an hour, were able to raise them to 3s. 6d. an hour when they arrived – this made all the difference to their morale and they worked happily. They had good camps and good food, and somehow they got a team spirit that enabled them to work ten hours a day and overtime too.'

King George VI occasionally drove over from Sandringham to Snetterton, Norfolk, where Taylor Woodrow were working on a large aerodrome. He used to inquire anxiously: 'Are you coming any closer to my boundary?'

The sub-agent at Snetterton was Tom Freakley, now director of Taylor Woodrow Ltd. and deputy chairman of Taylor Woodrow Construction. He had begun his career at fourteen in a Staffordshire coal-mine, hitch-hiked to London with 32s. in his pocket, run his own business as a brickwork sub-contractor 'on the tape', and joined Taylor Woodrow in 1940 as general foreman on the Innesworth Lane camp at Gloucester.

Chief engineer at Snetterton was Barton Higgs, now director of Taylor Woodrow Construction and deputy chairman and managing director of Myton Ltd. He had joined in August 1940, and after a spell on anti-invasion coastal defence at Portsmouth worked on aerodrome construction. In 1944 he took his invaluable knowledge of airfields into the R.A.F., and, as Flt-Lt Higgs, followed the Allied armies across Belgium, Holland, and Germany, repairing bombed airfields for fighters and bombers flying in close support of the ground troops. Three days before V.E.-Day, at Lübeck, he took the surrender of a swarm of German soldiers who had been left behind the retreat to destroy aircraft and installations.

Frank Pursglove was now in the Navy. He had been re-

1. Frank Taylor, chairman and founder of Taylor Woodrow.

2. (a) Réconstruction of Euston rail terminal.

2. (b) Wylfa nuclear power station, Anglesey.

placed by a sixteen-year-old girl, Christine Hughes, just out of Gregg's College. Her first job in February 1940 was as a junior shorthand-typist in the Estimating Department at Southall, where she gained a good basic training in the technical terms and language (not, of course, the site language) of the building industry. She proved her ability to stand the pace of the chairman's office by her mastery of the Ediphone machine, whose records she transcribed with great efficiency.

Christine, who some fifteen years later became Mrs Frank Taylor, now in these war years found herself driving the chairman all over the country to visit the company's contracts. These thousands of miles were covered mostly in a Ford 10. Her visits to aerodromes, ordnance factories, and camps were made in the blackout, air raids, bombs, and falling shrapnel from anti-aircraft guns. One night the shrapnel was so terrible that she had to dive under a lorry loaded with ballast. (Many Taylor Woodrow people found that a lorry full of soft ballast was far better protection than an air-raid shelter. The boardroom table, stoutly constructed, also came in useful for diving under when the doodlebugs came.)

The daily routine was this: she drove from Head Office while Frank Taylor read the mail and other papers and made 'action' notes on them. Then, while he was inspecting the contract on arrival, she typed up his notes and cleared as many matters as possible by telephone.

Then Frank Taylor would take the wheel of the Ford 10, dictating to her the points which had arisen during his visit to the works. If they were going on to another contract, she would type and telephone as before. If they were returning to Head Office, the procedure was adapted accordingly.

To this day, Christine Taylor never goes out to a site without slacks and rubber boots. This is because of an experience at Yelland, North Devon, when her boss asked her to go back to the office for some information and she tried to take a short

D

cut across the site. She was half-way across when she began to sink in the mud. Soon it was up to her waist. Luckily a foreman was near by and managed to throw a rope and pull her out. It was the middle of winter and there was only cold water in the site washroom.

'But it was fun,' she says today. 'We were young, we were ready to do anything and everything.'

One of the top-secret tasks Taylor Woodrow were given was the building of duplicate offices in Malvern for the evacuation of the Admiralty in case Whitehall was destroyed. It was also to become the headquarters of radar research. Hardly had the contract been started when the Admiralty asked: 'How much extra would it cost to complete the job in six months instead of nine?'

Frank Taylor said, '£20,000.'

The Admiralty hummed and ha-ed. 'How much to complete in three months?'

'£40,000.'

More humming and ha-ing; then – 'Well, we'll have to think about it. I'll let you know.'

They thought about it for three months, and then said, 'O.K. – in three months for £40,000. Can you really do it?'

'Do it?' Frank laughed. 'We've done it – it's ready!'

The general foreman on the Malvern project was Harold McCue, who had joined Taylor Woodrow in 1940 ('on All Fools' Day') from the building department of J. Lyons & Co. The son of a Sunderland and master builder, he had learned each trade the hard way – bricklaying and plastering (by tradition they are one and the same trade in the north), joinery, carpentry, and plumbing. Today he is director of Taylor Woodrow Ltd. and chairman of Taylor Woodrow Industrial Estates Ltd., Taylor Woodrow (Building Exports) Ltd., and Taylor Woodrow (Arcon).

•

Wherever there is top secrecy, there are ludicrous leaks in security. One day, almost as soon as he was put in charge of the Ministry of Aircraft Production, Lord Beaverbrook sent for Frank Taylor. He was very dramatic and mysterious. 'Have you men you can trust?' he asked.

Thinking of his thousand talkative Irishmen, and thinking also that he could successfully appeal to their discretion, Frank Taylor said yes.

The main task was to convert all the buildings in a requisitioned area at Brooklands, Surrey, into factories and workshops for making aircraft parts. This was to be a 'package deal', for Taylor Woodrow were to be responsible for everything, including the plans. But there was a still more secret project – a factory at Egham for assembling the aircraft parts once they had been manufactured. This was the private airfield used by Edward VIII when Prince of Wales, and planes were to be assembled here so that they could be flown away for their first test. A veteran illustrator called Norman Wilkinson had been engaged to look after camouflage, for it was perilously near Windsor Castle.

Beaverbrook lowered his voice to a whisper (and there are not many men who have heard the Beaver whisper): 'The code name – don't write it down, just remember it – it is VAX. 1. The project must *never* be referred to by any other name.'

A day or two later, Frank Taylor received a letter by ordinary mail addressed to him at 'The Aircraft Assembly Plant, Wick, Windsor Great Park, c/o VAX. 1, Egham'. And a couple of weeks after that, the King and Queen drove through Windsor Great Park to the site, and said: 'So this is the little hut Beaverbrook wanted!'

'VAX. 1' was only a small part of a complete dispersal of the Vickers-Armstrong aircraft factory at Weybridge to minimize the risks of bombing. This involved the erection, in only a few months, of thirty-two factory and ancillary buildings in five different counties.

The attack on Pearl Harbor brought America into the war. Among the first arrivals were units of the United States Air Force. By now Taylor Woodrow had built dozens of airfields and had all the experience anyone could want. But some units wanted to do their own building.

'The trouble was,' Frank Taylor recalls, 'that they had not got the materials. Sir Ernest Holloway, at the Air Ministry, asked us if we could help. I suggested a single contract, a package deal, by which we would supply *all* the American units. Sand, ballast, concrete, anything they needed – by this time we were organized so that we could do it. Sir Ernest agreed – and the Yanks got all they wanted. We never kept them waiting.' In two years, thirty-five American airfields were supplied with materials and technical services – and general manager of this, the company's first venture as agents on the grand scale, was Harold McCue, with head-quarters in Cambridge.

IV

While Bombs Fell

EVER since the tide of Britain's fortunes appeared to have turned, first with Hitler's suicidal attack on Russia, and then with Montgomery's victory at El Alamein which (in the words of the famous *communiqué* of October 1942) sent Rommel 'motoring westwards in top gear', there had been agitation, both at home and among our Allies, for 'a Second Front – Now'.

Guy Eden, a veteran lobby correspondent who had kept many official secrets during the war, afterwards revealed some of the reasons why the Second Front could not come any earlier than it did. In his book *Portrait of Churchill* he says:

'Strange little sideshows had been noticed by attentive readers of the newspapers, dating back as far as 1941. The island of Lofoten, Norway, was raided early in that year and a fish oil factory was destroyed. There were also raids of which the public heard nothing. They proved that it would not be profitable to raid the Continent on a small scale, that only overwhelming force would succeed.' Meanwhile, German propaganda boasted of the impregnability of the West Wall.

Churchill read all the reports of coastal raids and dictated the following directive:

Piers for Use on Beaches

C.C.O. or deputy.
They must float up and down with the tide.
The anchor problem must be mastered. Let me
have the best solution worked out. Don't argue
the matter. The difficulties will argue for
themselves.

W.S.C.
30.5.42

This was the seed from which the Mulberry Ports grew. Two were to be designed, one to handle 5,000 tons and the other 7,000 tons a day. Each was to be the size of Dover Harbour, which took seven years to build – and the time available was six months. Also they had to be built in secret. There were to be fifteen miles of piers, causeways, and breakwaters, weighing altogether three million tons.

'Concrete caissons, each as big as a block of flats, 146 of them, were to be built. In fields and all sorts of odd places, men got to work, building bits and pieces of the great ports without knowing what the strange-looking things were for.'

Several contractors were collaborating on this project, among them Pauling & Co., who were making 'Whales' (floating jetties linking pier-heads with the beach). Their boss, Sir John Gibson, was director-general of Ordnance factories at the Ministry of Supply, and all the contractors used to meet in his office. His senior civil engineer (Thames Area) was George Hazell, who was to join Taylor Woodrow in 1945 and is now managing director of Taylor Woodrow International.

A chargehand carpenter named Tom Reeves had just joined Taylor Woodrow at Barking, and was promoted to foreman carpenter in a few weeks. Today he is a director of

Taylor Woodrow Construction Ltd. A Yorkshireman who began his career at fourteen as tea-boy in a builder's yard in Acton, he was agent in charge of a million-pound contract at London Airport only eight years after joining in 1943.

Taylor Woodrow's contribution to the Mulberry Ports (or 'Operation Phoenix' as it was called) can best be expressed in figures: 120,000 cubic yards of excavation; 15,000 cubic yards of hardcore, in roads and groynes; 25,500 cubic yards of concrete; 85,000 yards of shuttering; and 2,500 tons of steel reinforcement.

For six months Taylor Woodrow staff and men worked through bombs in the Thames Estuary area to produce some of the monolithic concrete units, each weighing 7,000 tons. Each had to be built in a few weeks partly constructed in dry dock and then towed away for completion so that the dock space could be used for making others.

'Each caisson was 204 feet by 56 feet by 60 feet high,' Tom Reeves recalls. 'We called one of them Vivienne after my daughter. When each was up to fourteen feet three inches, it was towed away to West India Docks. Everyone did everything in those days, no class distinction, no union trouble. We used to work thirty-six-hour shifts – in the dark when the raids were on. Three times we were put out of action by doodlebugs. It's a funny thing, but it was always the signing-on book that was missing after a raid!'

However, there is a breaking point for men working such long hours, and one Saturday they went on strike. 'Frank Taylor came to see how we were getting on – in a black Homburg hat and his best black overcoat. Soon he was pouring concrete with us, till his coat was white. I doubt if he was ever able to wear it again.' Tom meanwhile was making up for the shortage of night crane drivers by doing the job himself.

Then, from all over Britain, trains and lorries converged on

Selsey in Sussex and the various parts – code named Phoenix, Gooseberry, Bombardon, Whale – were fitted together, ready for D-Day when they would be towed across the Channel and the concrete caissons would be sunk to form breakwaters round the floating piers.

Meanwhile we were losing too many aircraft, especially bombers returning from raids on Europe, because of fog-bound airfields. Once again, a personal minute from Churchill began the research which eventually overcame the problem. It was to Geoffrey Lloyd, Secretary for Petroleum, on September 26 1942:

'It is of great importance to find means to dissipate fog at aerodromes so that aircraft can land safely. Let full experiments to this end be put in hand by the Petroleum Warfare Department with all expedition. They should be given every support.'

This was the birth of F.I.D.O. (Fog Investigation Disposal Operation), for whose successful development Lord Cherwell must take at least part of the credit.

In *Flames Over Britain* (Sampson Low, Marston & Co. Ltd.), Sir Donald Banks describes early experiments at the still unfinished Staines Reservoir, Moody Down near Winchester, and Langhurst in Sussex. At first coke braziers were tried, but eventually lines of petrol burners proved more effective.

The Taylor Woodrow team was assigned to airfields in Lincolnshire. Sub-contractors were responsible for the pipe work, which had to carry 1,200 tons of fuel an hour, while Taylor Woodrow were among the contractors who secured them with heat-resistant concrete.

The first R.A.F. stations to have F.I.D.O. were Lakenheath, near Cambridge, and Graveley, near Huntingdon. On July 17 1943, Geoffrey Lloyd was able to tell Churchill: 'There is great news of fog dispersal from Graveley Airfield: a Lancaster bomber successfully landed in visibility of only 200 yards.'

F.I.D.O. first entered battle on November 19 1943, when four Halifaxes returning from a raid on the Ruhr ran into a heavy fog (visibility 100 yards) and were brought safely to ground. Thereafter the 'Haigill Fido Burner' was used at most Bomber Command airfields, and eventually by the U.S. Navy in the Aleutian Islands. It was reckoned that F.I.D.O. saved 366 British aircraft all told.

The trouble was, F.I.D.O. needed an awful lot of fuel. In mid-winter, a throughput of 168,000 gallons per hour was consumed to clear the air within a gigantic frame of fire. Special sidings for petroleum trains were constructed on the nearest railway to each airfield, and large pipelines were laid across country to pump the fuel to the tankage near the air-strip.

This need for swift oil transportation was common to both F.I.D.O. and the coming invasion of Europe. Lord Mount-batten had been present at the Moody Down experiments, and casually asked Geoffrey Lloyd: 'Could you run a pipeline under the Channel when we invade? The beaches and anchorages are likely to be heavily bombarded and oil tankers are so scarce.'

Lloyd went to see Clifford Hartley, chief engineer of the Anglo-Iranian Oil Company, who remembered some experiments in Iran with small-bore high-pressure pipes, and suggested using hollow cable. This was tried across the Bristol Channel, from Swansea to Ilfracombe, with H.M.S. *Holdfast*, a converted London cargo boat, as cablelayer.

The 'Hais cable', as it was called, was followed by another system, the 'Hamel steel pipe', made in factories at Tilbury. This was the shared inspiration of H. A. Hammick, chief engineer of the Iraq Petroleum Co. and B. J. W. Ellis, chief engineer of the Burmah Oil Co. (Hammick + Ellis =

Hamel). It was tried out between Portsmouth and the Isle of Wight, with H.M.S. *Persephone* as pipelayer. Could flexible steel tubing be laid, like cable, without 'kinking'? It could.

P.L.U.T.O. (Pipe Lines Under the Ocean) was now ready for large scale development. Six special ships were built for pipelaying, known as 'Conundrums' (Cone-ended drums). On land, pipelines were buried to protect them against bombing. A rectangular 'grid' of them was laid over an area roughly defined by Liverpool–Avonmouth–Walton-on-Thames and the Midlands, with offshoots to East Anglia, Southampton, and Dungeness. The work was farmed out to a number of contractors, among them Taylor Woodrow who laid their section, three feet deep, across the Kentish orchards, led by Tom Freakley and supervised by A. J. Hill.

'We didn't know where we were, or what the pipes were for,' A. J. remembers. 'It was the plum season, in the Garden of England, where I live now; and we worked from dawn until there was no more light.'

Coastal pumping stations were set up at Sandown, Isle of Wight (for supplying Cherbourg), and Dungeness (for supplying Boulogne). At Shanklin, I.O.W., a huge 620,000-gallon fuel tank (called 'Toto') was constructed under an acre of camouflage netting. From this, oil was gravity-fed to two batteries of pumps at Sandown and Shanklin. The pumps were concealed by putting them in Victorian villas and hotels on the front which had been rebuilt after bombing. At Sandown, an old fort, which had originally been built against the threat of an invasion from Napoleon III, was used. The whole Sandown–Shanklin installation was code-named 'Bambi'.

The secondary pumping station at Dungeness was code-named 'Dumbo'. This was in a much more dangerous position, being within range of German guns at Cap Gris Nez. Pumping machinery was hidden in little bungalows with names like 'Mon Repos', 'Sans Souci', and 'Y-Wurri', and it was supplied by underground pipelines from Walton-on-Thames and the Isle of Grain. At the Boulogne end (known as 'Dumbo Far') the pipeline emerged near the Imperial Hotel.

The first undersea pipeline between Sandown and Cherbourg was laid by H.M.S. *Latimer* in August 1944, and was first used on November 22. Soon 56,000 gallons of fuel a day at pressures of 750–1,200 pounds per square inch were rushing across the Channel to the Allied forces, and eventually petrol was going from the Mersey to the Rhine. In all, seventeen pipelines were laid, eleven of them Hais cable; and in its lifetime, P.L.U.T.O. pumped 173 million gallons of fuel across the Channel.

Taylor Woodrow might have borrowed the Royal Artillery motto *Ubique*, so various was their war work. Royal Ordnance factories, penicillin factories at Barnard Castle and Ulverston, with, in charge respectively Tom Freakley and Charles Waggett, who is now vice-president of Taylor Woodrow of Canada Ltd., and more airfields. The Air Ministry had more than made up for its initial reluctance to award contracts by appreciative letters like this one in 1943:

'Your output on the contract constitutes a record for a civilian contract in this area.'

And in the Midlands and Wales, Taylor Woodrow were digging opencast coal for the fires of industry.

In all these activities, the group was reaping the benefit of having acquired, in 1942, the Greenham Plant Hire Co., the Cranford Sand & Ballast Co., and other interests to supply machinery and materials for their ever-expanding tasks.

The new managing director of Greenham, within the Taylor Woodrow group, was Ted Woolf, who had joined Taylor Woodrow Plant Co. two years before; in those days its headquarters was a lean-to shed 100 by 30 feet. Ted Woolf, who had been nine years with Greenham, now found himself boss of his old company.

'I simply wasn't used to management,' he smiles today. 'But Frank Taylor thought I could do it, and I learnt the hard way. We had to handle some tough characters – in wartime you couldn't pick and choose: we employed anyone with a pair of hands.'

Today Greenham have subsidiaries abroad and over sixty units. Ted Woolf, who had become a director of Taylor Woodrow Ltd. as well as chairman of Taylor Woodrow Plant Co. Ltd., retired in 1970 after thirty-nine years with Greenham and thirty with Taylor Woodrow.

V

Problems of Peace

WITH D-Day safely behind them, and a favourable prospect of victory, the Government began to prepare for peace. Thousands of people had been made homeless by bombing; thousands more would need homes – or at least temporary homes – as the Forces were demobilized.

In 1944 an architect named Edric Neel had persuaded Taylor Woodrow and other companies to form the Arcon group (Arcon is short for Architectural Consultants), and to put up money for research into new kinds of structures which would be adaptable and quick to assemble. Chief of the immediate objectives was to design a temporary house which eventually was known as 'prefabricated'.

Arcon Marks 1, 2, 3, and 4 were rejected in favour of Arcon 5, which the Ministry of Works accepted and ordered 43,000 at a cost, including organizing the manufacture, distribution, and erection, of £61 million. The staff, which grew to 6,000, also organized the erection of steel and aluminium houses manufactured by other companies. A letter of commendation from Mr Charles Key, Minister of Works in 1949, paid tribute to the 'successful outcome of this vast operation'. Harold McCue was general manager of the temporary housing project, and he was eventually joined, in April 1945, by George Dyter, who is now a director of the parent company and chairman of Taylor Woodrow Property Co. Ltd. George

Dyter, who had started in his father's 100-man building and decorating business in Hampstead, had been contract manager for T. F. Nash, building Army huts, and had also built British Restaurants for the Ministry of Food.

Designing and building 'prefabs' was pioneering work, for nothing like it had ever been attempted before. It was from Harold McCue's close working with Stewarts & Lloyds, Turners Asbestos, I.C.I., Williams & Williams, and the Arcon architects that the prefabrication technique was afterwards adapted to structures for export. This led eventually to the formation in 1949 of Taylor Woodrow (Building Exports) Ltd., which had a turnover of £600,000 in its first year and began to send its structures all over the world.

West Africa was a likely market – not only for Arcon, but for everything Taylor Woodrow had to offer. Frank Samuel of Unilever had urged Frank Taylor: 'We've got quite a lot to do in West Africa. Why don't you go to Nigeria, the Gold Coast, Sierra Leone?' Lord Cole, then chairman of Unilever, as guest of honour at the Taylor Woodrow Group Annual Prizegiving in 1968, recalled the long partnership between his United Africa Company and Taylor Woodrow in which 'we have never had a cross word at any time'. Turning to Frank he said: 'It was twenty-two years ago that we first met. You came in to see my chief, Frank Samuel, while I sat down at the side of the desk and you discussed the possibilities of a joint venture. Frank Samuel said to you: "Well, when we go into a joint venture we like to have the majority." And you said to him: "Well, Mr Samuel, as a matter of fact, so do we." From that, we decided to do a joint venture on a fifty-fifty basis.'

George Dyter, who had gone out to Nigeria just after the war as 'acting colony manager', suggested that Arcon structures could achieve world sales if steel (which was then 'export only') could be made available. Having suggested it, he had to do it; and he spent much of his life in the immediate post-war years making three-day air trips (that was the time it took in those days) selling Arcon structures across the world, especially around the Caribbean. So much so that he used to get cables from Frank Taylor saying: 'For Heaven's sake have a holiday in Miami on your way back.'

This was the beginning of a whole programme of 'tropical architecture', based on lightweight welded tubular steel frames whose walls, if needed, could be made of local materials. Standard industrial buildings – even churches – to Arcon designs are today to be found in over 120 countries.

At home, Taylor Woodrow, like other building and civil engineering firms, had plenty of 'war damage' work. 'Holes in the ground': these are necessary for all new buildings, but there were too many of them. Much of the demolition (or 'wrecking', as they say in America) achieved for free by the German Air Force could be used to good purpose if money, men, and materials were forthcoming.

It was a time of both frustration and hope; a time for energetic and adaptable industries that were ready to cope resourcefully with utterly changed conditions. New houses were needed, power stations, factories, bridges, and many essential undertakings on which work had to be suspended during the war. But however frustrating these first years of peace might seem, they were nothing compared to the chaos of industry, especially of the building industry, which the country had known back in 1921, when young Frank Taylor had built his two houses in Blackpool.

Yet houses were built, on the principle that 'housing should be provided out of savings by the people for the people'. After the war, at Hayes, their old stamping ground, Taylor Woodrow were building three-bedroom houses for £1,200 to £1,400 each – £200 less than Council houses.

Let us take stock of Taylor Woodrow as it was in 1945. Its

subsidiaries were Taylor Woodrow Construction, Taylor Woodrow Homes, Taylor Woodrow Plant Co., Builders Supply Co., County Homesteads, I. W. Properties, Greenham Plant Hire Co., Improved Macadams, Cranford Sand & Ballast, and three associate companies in the U.S.A. A formidable body of know-how, some of it unique. What would the strategy be? Which new markets would be probed first? And how far could business be consolidated and expanded in America?

Almost in spite of themselves, Taylor Woodrow had to look abroad for expansion, even for survival. 'After the war,' Frank Taylor said in a radio interview, 'it was very essential that our exports should be stepped up. In my opinion the building and civil engineering industry is the spearhead that could really help with exports.'

Three people were sent to East Africa: John Fenton, Bert Rigg, and Geoff St Barbe Connor, an Australian who had been persuaded to join by A. J. Hill. There they met Frank Samuel of Unilever who suggested they should consider starting in West Africa, 3,000 miles away! They returned with this advice, also that it was a little too early to start in the East.

Frank Taylor was in New York with Les Turner, Taylor Woodrow's top man in America, who had gone out there in 1937. America had a housing problem almost as acute as ours. Although they had had no bomb damage, there was a pressing need for houses for returning servicemen.

The average American wanted a 25 by 35 foot house in its own 50 by 100 foot plot, with a lawn in front but no fences or hedges, with hardwood floors and central heating; and for 10,000 dollars he was going to get it – in Queens County, New York, on Long Island, and at Lake Success, territories which had been familiar to Taylor Woodrow before the war. At Queens County estate, 1,632 apartment suites were going

3. Midland Links Motorway – the Ray Hall interchange.

4. Over to you! Liverpool Metropolitan Cathedral.

up on seventy-nine acres, costing fifteen million dollars, of which just under half had been supplied by a syndicate led by Taylor Woodrow. These suites, each of three to five rooms, were imaginatively designed for community living: there were built-in laundries, recreation rooms, a shopping centre, gardens, lawns, curved roads, landscaping, and of course car-parks. Rents were from twenty-two dollars per room per month.

At home, in January 1947, Howley Power Station, Warrington, was at last completed by John Biscombe and his team under A. J. Hill's direction, after over six years of war interruption. Arcon temporary houses were still going strong, and much credit must be given to Wally Bullock, chief demonstrator at Beddington, Surrey, distribution centre, who erected and dismantled his demonstration model about a hundred times a week. At Great Yarmouth, the biggest ever Arcon estate, Shrublands, was put up to house people from the bombed-out area of Yarmouth.

Growth and the return to peacetime conditions focused attention on the Company as a social as well as a business community, and on the need for flexible management. So 1947 saw both the opening of the Taywood Social Centre at Adrienne Avenue, Southall, in May, and the foundation of the Junior Board of Management (now the Management Development Board).

In the appalling winter of 1947, everyone lost time, money and tempers. At Pwll-Dhu opencast coal site in South Wales, the staff were snowed up for six weeks, losing altogether four months of effective work. One night, when snowdrifts were up to the telegraph wires, the site team slept on the canteen floor because they were completely cut off from the nearest town, Blaen-Avon. By radio-controlling the different sites and by superhuman effort, they somehow got production up to 1,000 tons a day and turned loss into profit.

This year saw the birth of Taylor Woodrow (East Africa) Ltd. The wonder of the age was the then famous, and now almost forgotten, Groundnuts Scheme. It was going to grow food for the rationed, hungry nation, it was going to provide work and civilized living for thousands of Africans. Later we shall see why it didn't; but for the moment Taylor Woodrow were preoccupied with a contract to build 127 miles of six-inch pipeline to bring fuel oil from the East African coast into the heart of what was jocularly called Nutland, at Mikindani, Tanganyika.

Even if it had not been so difficult to get shipping space to supply the men on the spot, the scheme was doomed almost from the start by official bungling. Taylor Woodrow stuck it for three years, during which press criticism of the scheme grew ominously. Noel Monks, in the *Daily Mail* of December 20 1949, was scathing about the Overseas Food Corporation, but full of praise for Taylor Woodrow.

Noli, in the Southern Provinces of what was then Tanganyika, seemed destined to be 'the biggest town in the south of Tanganyika' where 'only the tsetse fly reigns supreme'. Forest had been bulldozed and tree-dozed into submission. The Taylor Woodrow sawmill was ready to turn 100 trees a day into five furnished prefabricated houses, at a total cost of £1,600 each. The oil pipeline was making its way from the coast to Shell's inland depot (still not built); so was the railway from Mtwara, the deep-water harbour that would one day, in theory, carry the wealth of the Groundnuts Scheme to Britain.

A shopping centre, a school, a bank, a hospital, sports fields, a club, a mess, police lines, a green belt, tsetse fly control posts – everything had been thought of. Just one little area of doubt: the survey had not yet been completed. That survey might reveal that Noli was not after all the right place for a new township.

The journalists were not slow to see what was wrong. 'This is a groundnut town that was stillborn,' wrote Noel Monks. 'But 150 determined men, working for a private contracting company, have smacked life into it. It is the happiest, best organized outfit I have seen, though high-level dithering at one time threatened its extinction.'

Noli had once been thought of as the centre of the whole groundnuts area – until it was found that the soil was unsuitable. So the G.H.Q. was removed to Nachingwea, thirty-two miles to the south. That was why Taylor Woodrow were concentrating on timber at Noli. Hurried economies had already reduced the target to five prefabricated houses per *week* instead of per day.

At Nachingwea, everything could be delayed by a lack of spares: if a jeep broke a spring, it could be out of action for weeks. Here the Overseas Food Corporation, apparently unable to learn from experience, was in control. But Noli, said Noel Monks, 'is a private enterprise town. The men work for a private company. . . . You can't help gaining the impression that if the whole scheme had been left to private enterprise, as was Noli, there would have been less bungling, squabbling and waste of money.'

Finally Taylor Woodrow collectively blew their top. On January 24 1950, Frank Taylor asked the Overseas Food Corporation to release his company from their contracts. 'Neither directors, executives nor staff have any confidence in the policy and ultimate success of the Groundnut Scheme' (under its present form of management, that is). 'Millions of pounds are still being wasted. After three years of experience, the Overseas Food Corporation is still making the same mistakes over again.'

He listed a number of basic things the Corporation should do, without which 'the scheme will not have a cat-in-hell's chance of succeeding'. There should be a hydro-electric dam

for irrigation and drinking water; cheap electricity for saw-
mills, factories, and lighting; irrigated gardens for growing
food instead of importing expensive canned food; better
roads, to avoid wear and tear on vehicles; the breaking
down of the whole project into small, manageable sections.
In a word, the broad, expert, imaginative 'package deal'
planning for which Taylor Woodrow have since become
famous.

'One of the chief reasons we are getting out,' Frank Taylor
concluded in a rare outburst of anger, 'is because we have
been mucked around.'

At home, and in other countries, the group's progress can
perhaps be measured by sitting in on the Taylor Woodrow
annual dinner on December 14 1948. One of the guests, the
editor of the *Contractors' Record*, commented: 'It is refreshing
to learn that some of our great building and civil engineering
firms have not waited for a full programme of contracts to
mature at home.' Or, he might have added, for Government
red tape to strangle effort.

'You may like to know,' Frank Taylor told the guests, 'that
we are contributing to the housing of people in England,
Africa and America. This year in civilized New York we are
well on the way to completing nearly 1000 beautiful homes.
In contrast, half-way across the world from there, in the wilds
of East Africa, at a place sixty miles from the nearest town,
we are providing comfortable tropical houses for the people
who win the soda from that extraordinary phenomenon, the
Magadi Soda Lake.'

Schools were being built as far apart as Nairobi and Ching-
ford, Essex; another penicillin factory for Glaxo was going up
at Barnard Castle, and factories for Scottish Industrial Estates
in what had once been a 'depressed area'. In Derbyshire, and
at Pwll-Dhu, the 1,500-foot mountain top in Monmouth-
shire, Taylor Woodrow were producing opencast coal (over

a million tons at Pwll-Dhu). Some of it would feed the West Ham power station, where they were also working.

'In South Africa we are constructing sewers and building houses. In West Africa, in a fifty-fifty partnership with the United Africa Company, a subsidiary of Unilever, we are building flats, houses, offices, warehouses, stores, motor depots and showrooms, and even a brewery.' (There were to be several more breweries in West Africa.)

Bob Aldred, now chairman of Taylor Woodrow International, was sharing a hut in an African village with Tom Freakley. Bob Aldred had spent most of the war in Sierra Leone, working for the Admiralty Works Department at Freetown dockyard. He had promised himself that once the war was over he would never set foot in West Africa again. Back in England in 1948, he answered a Taylor Woodrow advertisement – and in no time at all found himself back in West Africa, based mainly on Lagos. (After many more journeys there and elsewhere, for national bodies as well as for Taylor Woodrow, he also found himself appointed in 1970 chairman of the Export Group for the British Constructional Industries, a position which had been held by Frank Taylor from 1954–55.)

Tom Freakley, who had gone out to West Africa as agent in 1946, subsequently became colony manager, Nigeria. He recalls Taywood's first house, the price of cement (three times as much as in the U.K.), employing masons at 3s. 6d. a day, and the difficulty of teaching local bricklayers to use more than one kind of trowel. Could they be trained to lay 400 bricks a day? They could indeed: soon they were laying 150 bricks *an hour*. 'And we had to make our own bricks and floor-blocks,' Tom Freakley remembers. 'There were no local manufacturers then.' Tom, who is anything but a 'big white chief', believing that West African workers must be coaxed, not bullied, found that they learnt to drive machines easily, but were slower to acquire manual skills. 'It was a question of values,' he says. 'They valued time more than money.'

They were also getting trade-union-conscious, and when a strike got out of hand, Tom called in the police. For this he was attacked in the West African press: 'We pray for Mr Freakley's removal – some of our workers have received sudden boot from European massa.' In the long run, Tom's fairness and firmness paid off: metaphorical 'sudden boot' was no longer needed.

At Apapa the most modern joinery works in Nigeria had been built, to work in West African hardwoods such as iroko, opepe, agba, walnut, and many kinds of mahogany. They made every kind of woodwork for houses, and furniture for their own and other European companies' houses. And they discovered how good African craftsmanship in wood can be. The main office block housed Bob Hirst, Roy Wykes (now president of Taylor Woodrow of Canada Ltd. and of Monarch Investments Ltd.), and all the administrative staff.

On the then Gold Coast, quarrying was beginning for a new contract – the Takoradi Harbour extension. Ron Copleston, now director of Taylor Woodrow Construction Ltd., wrote a description of life in this bit of West Africa in 1950:

'Our staff accommodation is located in two main sections. Taywood Cove consists of a number of bungalows previously occupied by Naval personnel and built in a U-shape, the open part being seashore where our people bathe, and a new section, Windy Ridge, on high ground. At both Taywood Cove and Windy Ridge accommodation is provided for married and single people. The bungalows are fitted with mosquito-proof netting, and all are complete with small flower gardens.'

Just inside Taywood Cove was the Taywood Club, with table tennis, darts, radiogram, books, and billiard table.

Travel was mostly by small eight-seater Dove aircraft provided by the local air service.

This is fairly typical of the way in which Taylor Woodrow strive to make life comfortable for their staff, both British and native, overseas. Takoradi also provided an instance of another problem frequently encountered – the importance of taking local religions seriously. Some of the African workers were frightened of a spectre which, it was rumoured, had been sent by the gods to stop work on the harbour. So the correct steps were taken, and a local newspaper reported with satisfaction that 'a bullock was slaughtered for libation on the order of the fetish priests and priestesses of Sekondi on Saturday, October 28th'.

In London, Taylor Woodrow had been entrusted with the City of London's first major piece of reconstruction since the bombing – Atlantic House, a large office block designed by Sir Thomas Bennett to accommodate 1,500 people. This was Taylor Woodrow's first major office block, and Tom Freakley (assisted by Wally Higgs, now in Nottingham) had arrived back from West Africa to take charge of it.

The company was also applying specialized skills to 'the largest private conversion scheme ever undertaken' – Eaton Square, where Jim Skinner and his men were turning twenty-six elegant houses, originally built by the great Thomas Cubitt, into 103 flats and maisonettes, preserving the original façade. Still in the West End, the company were building a £400,000 block of flats for the Duke of Westminster and some more flats for the Metropolitan Police. Looking at all this evidence of the flexibility of private enterprise and its ability to break itself down into small units when necessary, it is odd to reflect that in this same year, 1950, there was talk of the nationalization of 'all building firms employing more than 20 men'.

As if to prove the folly of this, Taylor Woodrow (as

reported in the *Sphere* of March 25 1950) were carrying out an exercise at Cranford Park Estate which showed that twenty-five men could build a pair of houses in twenty-five and a half days.

In Oxford, over the river Cherwell in the leafy shades remembered by generations of undergraduates in punts, a delicate, almost fragile-looking footbridge had appeared. Its technical importance was out of all proportion to its size; for this was the first full-scale, single-span prestressed concrete, fixed-arch bridge, constructed by Taylor Woodrow. In charge of it was a promising young engineer called John Ballinger, whose rapid rise to top responsibility we shall see in later chapters. Soon there was another at Shrewsbury, Castle Bridge, the first prestressed concrete, balanced cantilever bridge in the U.K. and it also marks the debut of Dudley Harris, now chairman of Terresearch Ltd.

We will end this present chapter with a quick round-up of activities in 1952. Taylor Woodrow (Australia) Pty. Ltd. had been formed in 1951. There was not much business yet, but they were building the thirty-six-mile Geelong–Newport pipeline for Shell. In East and West Africa, there was still much to do, and the completion of Ralli House in Mombasa is but one example. Arcon tropical structures were doing well all over the world, and a thirty-foot span tropical roof was the mainstay of the 'Church of the Open Bible' in Jamaica. In London, Battersea 'B' power station was under way, and a secondary modern school for girls had been built at Feltham. And at Heathrow, the first of the really big Taylor Woodrow tasks for this great airport was beginning – the tunnel connecting Bath Road with the central terminal area.

As we approach Coronation year, we find the world-wide expansion and variety of Taylor Woodrow's tasks are now so great that henceforth we shall separate them into 'home' and 'away'.

VI

The Fifties – at Home

'BUILDING for the Future' is a slogan much associated with Taylor Woodrow. It could be argued that nobody in his senses builds for the past, although the world is unfortunately replete with structures that seem to have been designed solely for present requirements, and motorways whose dimensions seem based upon a lamentable under-estimate of the volume of traffic they will have to bear in twenty, or even in ten years' time.

Occasionally a construction firm will 'build for the future' by adapting old buildings of historic importance to the needs of modern living; we have seen this done in Eaton Square, and it was to be done again in Park Crescent in 1963, when, behind a replica of the original historic Nash Terraces, Britain's first International Students' House was built, giving residential accommodation for 135 students, and social, dining, and recreational facilities for 2,000 non-resident students.

This phrase, this 'Taylor Woodrowism', is worth putting to the test. In order to see how far the group were thinking ahead, say twenty years ago, let us look back to the early 1950s. There are good reasons for choosing 1953 as a point at which to pick up the story.

In that year began Taylor Woodrow's involvement, ahead of their competitors, on the civil engineering and building contracts for the world's first full-scale nuclear power station at Calder Hall, which was due to become operational three years later. This was a logical consequence of the group's association, for the last ten years, in power station work for the Central Electricity Generating Board.

Nineteen hundred and fifty-three also found them engaged in two other regular lines of business, aerodrome and sea defence works. They were erecting the subways, passenger-handling buildings, and nine-storey air traffic control building at London Airport that are so familiar to us today; but they also found themselves coming to the rescue, together with the Army, of the area stricken by the East Coast floods in January, reconstructing the damaged defences.

In his statement to shareholders, Frank Taylor took pride in the group's contribution to the Government's housing plan, which had exceeded its target of 300,000 houses. The easing of the licensing situation had helped to make this possible, and Taylor Woodrow were building three-bed-roomed houses, at prices ranging from £1,615 to £1,950, at Kidlington (Oxfordshire), Walsall, Burgess Hill (Sussex), Stowmarket (Suffolk), and High Wycombe. Prices included road and legal charges. The houses were offered freehold, most of the sites had room for a garage (this was before the mass ownership of cars), and with a ten per cent deposit the purchaser could repay the rest at 35s. to 45s. a week.

The secret was 'private enterprise', a point that Frank Taylor did not fail to rub in. In America, the Government had sought the help of free enterprise builders in its slum clearance programme, especially in the City of New York. By selling the land to private builders who designed and built apart-ments for letting, or built blocks of flats under a co-operative apartment development scheme, it was possible for people to buy a flat on deposit with repayments spread over thirty years. Why should not the same be done in Britain?

A Taylor Woodrow Annual Report is never a mere balance sheet with a list of board retirements and offers for reelection. It is a platform for Frank Taylor's views on the world in general. In 1953, he drew attention to the benefits given by the building and civil engineering industry, not only to the country in which the development takes place, but also to the export market generally. 'Directly a major civil engineering project is under way, extensive calls are made for steel, cement, paint, and the 101 items required to complete the task.'

Moreover, the employment of local labour increases the spending power of the people, the standard of living is raised, and so we have new overseas markets for consumer goods. The commercial risk is borne by the contractor, and here Frank Taylor was speaking on behalf of the whole industry: 'I am becoming increasingly aware of competition from Continental contractors who are making bids . . . for works at prices that we know to be well below the true value of the costs. This in my opinion is made possible by help from their Governments by way of tax relief or other concessions.' If British contractors were to compete in export markets, they must have an arrangement of this kind.

Nineteen hundred and fifty-four found the famous red and white Taylor Woodrow signs in every continent, at over 250 projects. In Perthshire, work began on a concrete buttress dam for the North of Scotland Hydro-Electric Board. In the Lowlands, on a new site, men and machines continued open-cast mining at Halkerston, where they expected to dig 580,000 tons of coal for the National Coal Board. In Birmingham a seven-storey office for the Prudential Assurance Company was going up, and in Reading, a new £500,000 building for the Faculty of Letters at Reading University. At Kirkby, in Liverpool, the first all-welded tubular steel frame factory (for Tubewrights Ltd.) to be wholly fabricated on site had been completed. And on New Year's Day, 1954, Taylor Woodrow entered upon its eighth contract for the Ministry

of Transport and Civil Aviation at London Airport, where it
had been given the gigantic task of building the central
terminal area; this time the job was the three-storey eastern
apex building – 'The Queen's Building'.

At home – Taylor Woodrow's own home – the new head
office at Southall, Middlesex, was completed and opened on
June 1 1954. It had been built in fourteen months – a notable
achievement, considering that the existing offices had to be
kept open during construction. It was carried out in three
phases, so that new space was provided before the old accommodation was destroyed. When Taylor Woodrow's
client is Taylor Woodrow, exactly the same planning and
scheduling methods are used, with the same aim of finishing
either dead on time, or well ahead of it.

The new building was 'open-plan', a novelty in Britain at
the time, though a commonplace in America. This arrangement gave maximum flexibility to alter the layout of departments as business demands, and avoid wastage of space. But
when there are no interior walls, the great problem is noise-
control, and this was attained by acoustic tiling. 'Open-plan'
also gives everyone more light, whether daylight or fluore-
scent lighting.

Another challenging project of 1954 was the start of work
on the Castle Donington power station. This huge contribution to the national 275,000–volt high-tension supergrid to
pool power resources was sited near the Leicestershire coal-
fields, to cut the rising costs of transport.

The Castle Donington project had begun on a rainy day in
October 1952 when the Taylor Woodrow advance guard
arrived to begin levelling and draining the site. 'Muck-
shifting' is the good old honourable name for this essential
exercise in the construction business. A third of a million
cubic yards of gravel, red mud, and green sandstone would
have to be shifted. As the site was three feet below flood

level, the team had first to build defences against the River Trent, which had so often swamped these fields.

'Soon the landscape was being remade,' wrote a contemporary observer, 'with giant excavators clawing up the ground, and steam hammers driving sheet piles into place; every day fleets of buses brought in the Taylor Woodrow army – 800 strong at its peak – from Derby, Nottingham and Leicester.' So many lorry-loads of gravel arrived from the nearby pits that it was estimated, after the job was completed, that Taylor Woodrow lorries had travelled 225,000 miles (a little less than the distance to the Moon) carrying material backwards and forwards between the pits and the site.

'Once upon a time,' the observer continued, 'the landmarks of Castle Donington were the twin spires of the Parish Church and the Methodist Church, crowning the hill above the Trent; but soon the traveller's eye will be caught by the two tall chimneys of the power station, rising to a height of 425 feet – three times taller than Nelson's column.'

Nineteen hundred and fifty-five was marked by two Royal occasions – the inauguration of London Airport by the Queen, and the inspection of Calder Hall atomic power station by the Duke of Edinburgh. Heathrow, described by Patrick O'Donovan of *The Observer* as 'the proudest thing that has been built in Britain since the war', gave rise to one of Frank Taylor's favourite 'true stories' – about the French lady who thought its name was 'Taylor Woodrow Airport'. Calder Hall was a double triumph, for the company had now been asked to build a second power station next door to it.

In this year, too, Taylor Woodrow purchased an interest in Myton Ltd., a Hull based building and civil engineering company which was to make an important contribution to the group.

London Airport, Calder Hall – what have these two vast undertakings in common? The use of concrete, the preliminary 'muck-shifting', geological problems, the need to pump away thousands of gallons of unwanted water, and a hundred finer points that only engineers can appreciate. That is a strange thing about construction work: from each project, you learn something that, often unexpectedly, can be applied to a totally different project. Later, we shall see how the behaviour of concrete in nuclear reactors was studied in such a way as to provide a special kind of know-how for the building of a cathedral. Much that was learnt at London Airport came in useful at Calder Hall. Since the teams at each site were led, to a great extent, by the same men, the exchange of information was not difficult.

The director-in-charge of London Airport central terminal buildings was Tom Freakley, with Tom Reeves in charge of civil engineering: and John Ballinger, whom we first met at the Oxford footbridge, was agent for most of the tunnel contract. Two of these, Messrs Reeves and Ballinger, were also in charge of Calder Hall, the latter resident on site as agent and subsequently as contracts manager.

Calder Hall was opened by the Queen, who in October 1956 threw the first switch to feed electricity produced by atomic energy into the national grid. A. J. Hill, Tom Reeves, and John Ballinger represented Taylor Woodrow at the ceremony. In the same year, the company gained two more power station contracts, High Marnham and Northfleet, bringing to a total of fourteen the number of power stations on which they had been engaged so far. They were also providing some of the fuel for these stations, for they were digging more opencast coal than at any time since they had begun work in this field in 1942.

Everybody enjoys watching other people at work. Frank Taylor had noticed how Americans treated rubbernecks, who were after all members of the public, and might be even shareholders or prospective clients. If they wanted to watch, why not make them comfortable, so they did not cause a pavement obstruction? So he instituted Britain's first 'public observation platform' so that people could see what was going on behind the hoarding. It first appeared at Twenty Albert Embankment, on the south side of the Thames, and afterwards at a seventeen-storey office block, Bowater House, Knightsbridge. *Punch* seized upon it in a cartoon showing people queueing for the platform, which in fact they did.

At Southall, new laboratories had been built and equipped, to enable the company to improve and develop in concrete technology, ground exploration, soil mechanics, chemistry, and metallurgy, and to test building materials of every kind. These were designed by Reg Taylor, director responsible for research.

The Arcon organization showed its highest ever turnover during 1956, and prospects for further expansion seemed excellent, although the rise in the cost of sea freights following the closing of the Suez Canal affected its ability to compete in certain foreign markets. Greenham's plant-hire facilities were also extended by the opening of new premises in Manchester, so that up-to-date machines were always available, not only to the group's own construction companies but also to the many clients who used this hiring service.

It had begun in 1930, when Nick Greenham, owner of a sand and gravel pit at Isleworth, Middlesex, first let out on hire a dragline excavator, and today it is a nationwide organization, supplying a twenty-four-hour service from Scotland, North Wales, the North-West, North-East, and the Midlands.

Meanwhile, the Lednock Dam in the Scottish Highlands was nearing completion. This had involved hauling heavy plant and machinery up six and a half miles of mountain road to a height of 1,000 feet; drilling, blasting, and excavating rock; and diverting the river. Living-quarters for 250 men had to be provided. The camp had all the domestic and recreational facilities needed for a self-contained community that would have to endure a couple of severe winters at a considerable altitude – with its own hospital and trained medical staff, a sewage works, a shop, a post office, a bar – and a priest to conduct a service every week. Frank Taylor's son-in-law Brian Trafford, as a young engineer, and his family were among many others in the team who moved to Scotland to work on this contract.

Two gaps had to be left in the dam for river diversion and an access road before the prestressed bridge along the top could be installed, and these were nicknamed 'The Toothless Monarch of Lednock' by the locals.

Nineteen hundred and fifty-six saw the coming of age of Taylor Woodrow as a public company, and also the twenty-fifth anniversary of the move south from Lancashire. Although the group was now undertaking vast engineering and constructional feats all over the world, it was still conscious of its beginnings in ordinary housebuilding. Taylor Woodrow Homes was now building 3,500 houses on nineteen major estates, some of them conventional 'semis', others designed on modern 'open planning'. At Luton, Harlow, Ashford, Crawley, and other places, the company was offering houses at £1,895 to £2,350, and luxury homes from £2,495, including land, road charges and services, trees and landscaping. No wonder people were willing to spend the night in parked cars to be first in the queue to reserve a plot.

In Somerset, a six-mile road had been built – for a special purpose. It was to be the access road to Hinkley Point, the future 'world's first' 500-megawatt nuclear power station, which the English Electric-Babcock and Wilcox-Taylor Woodrow Atomic Power Construction Company had designed in association with the Central Electricity Generating Board. The road, like so many of the group's projects, had

9(a). Greenham on the Warspite. Compressors help in raising the famous old battleship from the rocks of Prussia Cove, Cornwall, where she went aground while being towed to the breaker's yard.

9(b). Public Observation Platform with a difference. Closed circuit television at the Midland Bank site in Gracechurch Street.

10(a). The Monarch of Glen Lednock after teething!

10(b). Observation platform in reverse. The construction team in Victoria Street look on, as H.M. Queen Elizabeth II accompanies President Abbaud of the Sudan to Buckingham Palace on a State visit.

11(a). Man-made valleys. Opencast coal production in Wales.

11(b). Commerce in Nigeria. Extensions to double the capacity of Port Harcourt.

12(a). Hinkley Point. Bill Johnson, agent, has a visit from (*left to right*:) Frank Gibb, Frank Taylor, John Ballinger, and A. J. Hill.

12(b). Teamwork. Standing beside the 100-ton block of Cornish granite from which he carved the largest monolithic sculpture of modern times, David Wynne is seen here with his preliminary model.

been finished ahead of schedule; and it is interesting to note that the director in charge of this (and of the main station contract) was John Ballinger, recently appointed to the Board of Taylor Woodrow Construction, with the rugged figure of Vic Blondell as agent.

Myton Ltd. were meanwhile working on B.P. House, an impressive office block in the City of London, and several other projects in London and the north. Taylor Woodrow Industrial Estates Ltd., recently formed to supply and erect industrial buildings on a leasehold rental basis, had already undertaken a pioneer project of this kind, in Scotland at Dixon's Blazes, thirty-four acres of former blast furnaces in the heart of the Gorbals. And Taylor Woodrow Construction Ltd. had carried to 'topping out' stage an extension to Sir Edwin Lutyens's Grosvenor House Hotel (where all the Company's annual general meetings are held). This was being done with special anti-noise precautions to avoid disturbing the guests; and so pleased were the chairman, Sir Charles Taylor, Jay Jones, managing director, and the Grosvenor House board with the fact that the extension was completed nine weeks ahead of schedule that they presented Taylor Woodrow with a £5,000 bonus cheque provided under the terms of contract.

In 1958 the Government decided to limit opencast coal-mining operations, and while the Company had enough work of this kind – two and a half million tons – to last two or three years, it was felt that there would be no immediate expansion to look forward to. In fact, later operations on the Royal Arms Group site in Wales alone have reached three and a quarter million tons to date. Charles Carr, agent there, who joined Taylor Woodrow in 1938, has been on this mountainous site since 1958.

In London, Bowater House had been completed for The Land Securities Investment Trust Ltd., of which Sir Harold

Samuel is chairman; and work was going ahead on the English Electric headquarters in Aldwych, on the derelict site of the old Gaiety Theatre. Schemes to modernize British Railways, missile bases for the Air Ministry, Harrod's new store (Rackhams) in Birmingham, buildings for Carreras, Cow & Gate, and the Wellcome Foundation . . . if this reads like a miscellaneous catalogue, it is simply that we have no space to tell the story of each, so full and so varied was the order book in this year. Symptomatic of expansion was the opening of Western House, Western Avenue, Ealing, the new headquarters for the overseas companies, Taylor Woodrow Homes, and Myton Ltd. A serious blow to the company at this time (late in 1959) had been the sudden death of Peter Elliott, director and general manager of the Myton company, still a relatively new member of the Taylor Woodrow group. Barton Higgs was moved over at very short notice from Taylor Woodrow Construction to take his place.

Competition was hotting up in the construction industry. Bank Rate had risen, and there was much talk of restricted credit. Coming up to the silver jubilee of its incorporation as a public company, Taylor Woodrow showed record profits for 1959. This, of course, was 'not unsatisfactory', Frank Taylor told shareholders; what *was* unsatisfactory was that 'severe competition' was expressing itself in the behaviour of certain companies over tendering: 'a contractor is bound to produce the work according to the specification and up to the quality required, therefore he is entitled to have a fair price to cover the actual cost and a reasonable reward for his labour and "know how"'. Yet 'in every case where tenders are called for, it would appear that one or two contractors are going in at figures well below cost'.

Prospective clients, he went on, might like to consider Taylor Woodrow's 'turnkey' form of contract; 'we have our own people together with other specialists such as architects, consulting engineers, and quantity surveyors who form a team which can plan, design, estimate and complete the project from drawing board to nameplate', the three watchwords being quality, economy, and speed of completion. In the decade that followed, the industry was to see a much greater degree of co-ordination and large-scale teamwork along the lines Frank Taylor advocated.

At Dungeness the Company was building its first lighthouse – indeed, the first lighthouse to be built in Britain for more than fifty years. It was also, with its associates English Electric and Babcock and Wilcox, tendering for another atomic power station next to the lighthouse (unsuccessfully, on this occasion) and was working on designs for yet another at Sizewell, Suffolk, which was to become the second to be carried out by the consortium.

A new hydro-electric project, comprising a dam, a power station, shafts, and tunnels, had just been started at Cwm Rheidol in Wales. At London Airport the Company was adding aircraft standings and taxiways to its other enormous works there; and off the Dorset Coast at Winfrith, despite heavy seas, a submarine pipeline for the U.K. Atomic Energy Authority, welded in association with the Collins company of Texas, was successfully launched.

The twenty-fifth anniversary of the company's 'going public' had highlighted the number of long-service team members. There were fifty-six members, in 1959, of the Taylor Woodrow Group '25' Club. Two of them were Bert Rigg, a director of the parent company since 1945 who retired as an executive director but retained his seat on the board (he did not take his full retirement until 1966); and Les Olorenshaw, already chairman, managing director, and director of several subsidiary companies, who now joined the main board after twenty-eight years' service.

VII

'Not Unsatisfactory'

TAYLOR WOODROW entered the sixties with caution. Nineteen hundred and sixty was a difficult year for building and civil engineering work at home, largely for an unpredictable reason: the weather. East of Bombay, you know pretty well where you are with the climate: the monsoon usually arrives in the first week of June, appears daily at about half past three in the afternoon, and departs about three months later.

But in Britain you never know. The weather was certainly to blame for restricted output of houses, and threatened to delay the hydro-electric project at Cwm Rheidol: that the works were, in fact, completed on schedule was greatly to the credit of the team concerned. Nevertheless Taylor Woodrow this year showed reduced profits despite increased turnover.

With English Electric and Babcock and Wilcox, its partners in the atomic power consortium, the Company successfully tendered for the new 580 MW nuclear power station at Sizewell, which was planned to be in operation by 1966. This was fortunate timing, for construction work at Hinkley Point had passed its peak and many of the men and machines which had acquitted themselves well at Hinkley could now be transferred to Sizewell.

London saw a great many Taylor Woodrow signboards this year. In the City there was a new headquarters for the

Overseas Branch of the Midland Bank. Myton Ltd. was responsible for Austral House, Kempson House, and other office developments, and attracted much public interest by their 464-car Zidpark automatic multi-storey garage near Southwark Bridge. There were Taylor Woodrow contracts for multi-storey office buildings in Victoria and Oxford Street and St John's Wood, for a large housing development for the Norwich Union.

In Birmingham, Tom Hyslop's team, under the overall direction of Tom Freakley, had completed Rackhams store. (Tom Hyslop is now a divisional director of Taylor Wood-row Construction.) It was designed to hold 10,000 shoppers at once, if necessary, and to include a highly sophisticated air-conditioning system, and also the shopping arcades which are much loved in Birmingham. One of the outstanding technical feats of this contract was the 'bridging' of the railway tunnel through which the Snow Hill–Paddington line runs underneath the site. Rackhams' new building was opened, nine months ahead of schedule, just in time for the Christmas 1960 trade.

Perhaps the most outstanding achievement of the following year was the Company's appointment as main contractors for the building of the New Metropolitan Cathedral of Christ the King at Liverpool, designed by Mr (afterwards Sir) Frederick Gibberd, C.B.E.

In contrast, the Company this year introduced a new line of homes, and their Sandown bungalow, shown at the New Homes Exhibition at the Central Hall, Westminster, was voted 'Home of the Year' in the up-to-£4,000 price range.

Twelve miles north of Doncaster, Europe's biggest viaduct of its kind, the 470 feet long prestressed concrete span at Wentbridge, had cut out one of the A.1's worst bottle-necks and restored a quiet village to normal peace. With its two pairs of raking legs, it struck one observer as 'a coffee-table for a modern giant' – a clean, contemporary design approved by the Royal Fine Art Commission. It is almost true to say it was built 'top to bottom', since the deck was well advanced before the leg supports were completed, a reversal of bridge-building practice made possible by a 'birdcage', involving 120 miles of steel scaffolding, strung out across the valley. The agent for this job was Bill Mangan, who has carried out many civil contracts in the north of England. The site team – over 200 at peak – had to combat the incessant rains of 1960 which flooded the valley, and exceptionally high winds.

Meanwhile, at Cwm Rheidol, Cardiganshire, rain and rivers had been harnessed to light and power British homes and industries. Taylor Woodrow's contribution to the then biggest hydro-electric scheme in Wales, directed by Tom Reeves with Joe Graham as agent, took place on three sites, and one of them twelve miles from the central office. It involved a dam to control the River Rheidol so that the River Board could reclaim land and train the river into a more stable course; the driving of a 2,000-feet concrete-lined tunnel; a power station in the Rheidol Valley; and a 'regulating reservoir' below the Rheidol Falls (near the famous beauty spot known as Devil's Bridge), to even out the flow from the station and prevent disturbance to the river banks and inconvenience to riverside dwellers. A 'fish ladder' was constructed alongside the Rheidol Falls (hitherto impassable to fish) to open out further reaches of spawning ground. Anti-pollution units on the higher reaches of the river were also provided.

Electricity, gas, coal – all three fuels figured in the group's work during 1962, a year which ended in chaotic conditions caused by the coldest winter for more than thirty years. High Marnham, then the first 1,000 M.W. conventionally powered electricity station in Europe, was opened, Taylor Woodrow being responsible for much of the civil work. Contracts to lay two major sections of the countryside network of methane pipelines were gained from regional gas boards, whose 325-mile project (from Essex to Yorkshire) was destined to bring natural gas, shipped from the Sahara Desert to Canvey Island, to householders and industry.

At Euston Station, Taylor Woodrow signs had gone up to announce the complete redevelopment of this main London terminus for the north. Myton were again in the news, with a £2 million office project (including a twenty-two storey tower which has become a major landmark) on the Kingston-By-Pass at Tolworth, Surrey, and two more major office blocks at Leeds City Station and Isleworth, Middlesex.

For some time, Myton, in conjunction with Anglian Building Products of Norfolk, had been studying techniques for making and erecting factory-made housing units. The result was the formation of Taylor Woodrow-Anglian Ltd., a company jointly owned by Anglian Building Products Ltd. and Myton, which now began work on a £2 million scheme for the then London County Council, to manufacture and erect 560 dwellings at Morris Walk, Woolwich. This was to be a forerunner of many other ventures of the kind.

Nineteen hundred and sixty-two was a busy year for Taylor Woodrow Industrial Estates Ltd. This property development company had embarked on a programme for the development of eight sites, one of which, Dixon's Blazes, we have noted already.

On October 28 John Hanson resigned from the main board through ill-health. A. E. Aldridge, who died in this year, had resigned the year before (both had been founder-members of the first public company in 1935). Reg Heasman became chairman of Myton, and Harold McCue, chairman of Taylor Woodrow (Building Exports) Ltd. and other companies. (Both had been appointed to the parent board in

1961.) Reg Heasman's place as parent company secretary was taken by Robson (Robin) Christie, another northerner – from Sunderland – who had joined Taylor Woodrow after the war, and became a director of Myton in 1965. A further change in 1962 had been the appointment as a non-executive member of the Taylor Woodrow parent board of Mr Charles Hambro, managing director of Hambros Bank Ltd.

A 1962 achievement worth examining in detail was the Ferodo factory opened by Princess Margaret on May 16. It created an entirely new industry at Caernarvon, North Wales, gave immediate employment to 500 people and eventually to 800 people. From moving the first earth to full completion of the brake-lining and motor accessory factory was sixteen months. This was an 'all-in service' contract, with Taylor Woodrow being responsible for everything, from the first briefing in October 1960 through fourteen stages – estimates, outline drawings, local authority and other approvals, excavation, steelwork, cladding, etc. An essential part of the planning was that it should harmonize with famous surroundings of great scenic beauty, and not interfere with the view of Snowdon from Anglesey.

Nineteen hundred and sixty-three showed record profits for the group, said Frank Taylor in his 'Dear Fellow Workers' letter at the end of his Annual Report, 'despite the hundreds of thousands of pounds lost during the worst winter of the century'. There was in fact a £50 million order book, £37 million of which was under *negotiated* contract.

'We have invented and are now marketing throughout the world a hydraulically operated *silent pile driver*. We have taken the United Kingdom licence for the S.A.F.E.G.E. monorail system which I believe can make a significant contribution to the relief of traffic congestion in our towns.' The Pilemaster and S.A.F.E.G.E. will be discussed in later chapters; they are both evidence of Taylor Woodrow's endless quest for an ordered environment amid progress and change.

From time to time land and properties with good investment opportunities had been offered to Taylor Woodrow, and the Company now felt the need to establish within the group a property investment company for buying and developing suitable land and properties to be retained for investment. So a new company, Taylor Woodrow Property Co. Ltd., came into being.

Another subsidiary, Terresearch Ltd. (whose managing director is Bill Calderwood), normally concerned with site investigation and foundation engineering, was launching its new 'Terrephragm' system for the construction of diaphragm walls in foundations.

It so happened that the theme of the Lord Mayor's Procession this year was 'The Building Industry', and Taylor Woodrow's float represented 'Building for Export', the float being preceded, as might be expected, by four men apparently towing it on a rope.

There was at this time a certain amount of criticism of the building and civil engineering industry, implying that there were areas of inefficiency in it. Co-ordination has always been the vital need, and there are sophisticated modern ways of securing this. By 1963 an I.B.M. 1401 computer installed at Southall in 1962 was speeding up costing information, accounts, and quantity surveyors' applications. It was also used by design and research staff, and for 'critical path planning' (getting the right quantity of materials and men to the right place at the right time) of the more complex contracts. It is particularly useful in the pre-planning stage; it is much cheaper to discover planning inaccuracies on a computer than to have to correct them in the field. The computer department is managed by Harry Froud under the direction of Martin Hebb, chief accountant at Southall, a long-serving

member and director of the services company since 1956.
Harry Froud joined the group in 1957 and has been with the
computer department from its inception.

A new depot for Fords at Thurrock, Essex; a factory for
Golden Wonder Crisps at Corby; work for Whitbread,
Debenham, Rio Tinto, Land Securities; tunnels for the
power station at Fawley, driven both into the Solent and
under Southampton Water; and civil works at West Burton,
where the first 2,000 M.W. power station in the country was
much admired by the Shah of Persia . . . the achievements of
the group at home in 1964 might seem manifold in this
year of record profit, despite a heavy loss on an overseas contract. But there was a clear statement of policy. With a
change of Government, economic difficulties and uncertainty
about future taxation, Frank Taylor told his shareholders,
'It is essential that we increase the volume of work undertaken
overseas, with its attendant risks, if we are to continue to
make a real and substantial contribution to exports.'

One promising export line was the Arcon Export House,
designed for housing projects in tropical and sub-tropical
countries, which Taylor Woodrow (Building Exports) Ltd.
were supplying to Pakistan, British Honduras, and two of the
loneliest places in the world, Ascension Island and St Helena.
Another prospect, it seemed, was the Pilemaster, seen (but not
heard) on various projects including the Hendon Motorway,
a Manchester office block, and a new bank in Belfast. The
silent, vibrationless pile driver, once people had been convinced that such a thing was possible, was already bringing in
inquiries from Europe and other continents.

Two years later, it was to win for Taylor Woodrow the
coveted new Queen's Award to Industry, 1966 – for technological innovation – and was by then on sale as far away as
Japan where it is manufactured under licence by Mitsubishi.
The Award was also gained by Taylor Woodrow (Building

Exports) Ltd. for 'export achievement', the outstanding example of which was a complete Arcon chicken farm designed and built for Hungary.

Another technical achievement, of a totally different kind, was the completion in 1965 of the Fawley transmission tunnel beneath Southampton Water with Colin McKillop as agent. Two miles long and ten feet in diameter, it was driven from opposite banks of the estuary simultaneously by men working in compressed air, and the link-up, when miner Joe McFadden and fitter Alex Quinn eventually shook hands across the rubble on October 29, was to an accuracy of one and a half inches! Colin McKillop, the group's tunnelling specialist, is now a divisional director of Taylor Woodrow Construction.

Among many other completed jobs in 1965 were buildings for Unilever and Bowater; the Tower Hill complex in the City, and multi-storey car parks at Taylor Woodrow's old stamping ground, Heathrow Airport. Two relatively small tasks may be picked out to show the company's more delicate skills: the main façade of International Students House in Park Crescent, Regents Park, opened by the Queen Mother on May 4; and a new chapel at Hopwood Hall Teachers' Training College, built for the Trustees of the De La Salle Order.

During the year, a new company, Taylor Woodrow Construction (Midlands) Ltd. was formed under the chairmanship of Tom Freakley, with headquarters in Stafford, to cope with the increasing volume of work in this growth area. From the start, this first regional subsidiary has been an outstanding success, thanks to men like Eddie Varcoe (deputy chairman) and Jim Millar (managing director), both of whom had been at Calder Hall. In its first year, it was carrying out major schemes for British Rail, Boots, Ford, and English Electric. Encouraged by this success, in 1966 a northern regional company was formed at Darlington, with Tom Reeves as chairman and Frank Carr as managing director, and immediately set to work on a £1½ million contract for runway works at Manchester Airport, among other important projects.

Back in 1965, however, with new Government restrictions on building, the group was having to fight harder for home growth. It was also, like other construction companies, having politically motivated labour difficulties. In his report to shareholders, Frank Taylor's sense of frustration appears:

'We have been subject to the activities of some disruptive elements which, although small in number, have a big effect and are no good to our industry. They are acting contrary to the real interests of the trade unions, they are flouting the authority of the industry, by provoking unofficial stoppages, time wasting and go-slow tactics. . . . Somehow we have got to find the answer to it.'

The worst of these troubles were experienced by Myton at the huge Barbican development in the City of London where Brian Trafford was director-in-charge. Things were to get worse before they improved. Better news for Myton in the sixties was progress on spectacular projects they had been selected to carry out in close association with the local authorities. Both were comprehensive residential, shopping, and office developments – among the largest the group had undertaken so far – one on fifteen acres overlooking the sea at Churchill Square, Brighton and the other on seventeen acres near the Clyde at Anderston Cross, Glasgow. A start had been made at Brighton in 1964 and at Glasgow in 1966; both projects are being completed in planned phases concluding in the seventies, and are being carried out in conjunction with the Standard Life Assurance Co. Also in conjunction with Standard Life is an extensive housing and office development in York Gate, Regent's Park, some lying behind two restored Nash façades. This was begun in 1967 (the year Myton opened their Glasgow Regional Office) and is also being completed in phases. Ted North is the Myton director responsible for this type of development, and is also on the board of Taylor Woodrow Property.

Meanwhile, Brian Trafford had made a film of rapid construction techniques at a nineteen-storey office block in Colliers Wood – structurally completed in fifteen weeks.

Taylor Woodrow Construction in 1965 started making a new contribution to the national roads programme – a £9½ million section of the vital Midland Links Motorway connecting the M.1, M.5, and M.6. There was a new nuclear contract, too, at the prototype fast reactor at Dounreay, Caithness. Sizewell, now finished, was opened by the Lord Lieutenant of Suffolk on April 7 1965. And, in many ways that most astonishing of all achievements, Liverpool Metropolitan Cathedral, was completed – we shall tell the full story of it in Chapter XIII.

In Italy, willing helpers were struggling to restore the art treasures damaged by the disastrous floods in Florence. Unexpectedly, the Greenham group of companies found themselves in a position to lend a hand – or rather some highly specialized machines. 'Drying out' equipment – Trembath industrial dehumidifiers – organized by Joe Gaffney, managing director of Greenham Equipments Ltd., were used at the Uffizi Gallery, having been flown to Italy by arrangement with the Italian Art and Archives Fund whose headquarters were at the National Gallery, London.

One subsidiary which has so far made only a brief appearance in this account is Swiftplan, who make a range of industrialized timber buildings and office partitioning. It reached record sales in 1966, and was awarded an appraisal certificate by the National Building Agency, after which a large housing contract was obtained from the Corby Development Corporation – eighty-eight Multiflex houses fronting on to a paved courtyard. In addition to the directors already named, managing director is Vernon Beck, who returned to England after many years in charge of Taylor Woodrow's West African joinery works. Taylor Woodrow was also one of its own customers, for the new Taywood Sports and Social Club at Southall used the Swiftplan 8042 system.

'Not unsatisfactory', the Taylor Woodrow family joke, appears in every Annual Report. So far as this historian can discover, it was first used in a Christmas message to staff in 1947. Twenty years later it graduated to inverted commas, and summed up a successful year, but a critical one for labour relations. Lord Cameron's Court of Inquiry on the Barbican situation vindicated the action of Myton. The unofficial pickets who had stationed themselves outside the site for many months against all union advice and warnings were withdrawn on November 2 1967. In a message to the group, Frank Taylor congratulated the Myton team 'under the untiring leadership of Mr Barton Higgs, together with the officials of the National Federation of Building Trades Employers and those responsible trade unionists and leaders who have brought this affair to its only logical conclusion'.

Meanwhile British Rail's £1.6 million extension to their technical centre at Derby had been opened and their office and computer centre at Crewe had been traditionally 'topped out', when 'Thirteen floors in six months is not to be sneezed at', was the comment of Mr A. N. Butland, British Rail's chief engineer. More good time-keeping resulted in the handing-over in December of the £2¼ million Daventry Parts Depot to the Ford Motor Company. It was completed in fifty weeks (two weeks ahead of schedule), and drew an enthusiastic letter from Sir Leonard Crossland, Ford's deputy chairman: 'This splendid achievement is an outstanding tribute to your organization and all its people. I would not presume to single out individuals but would rather say that the overall team effort was of an exceptionally high standard.' Also in 1969, a start was made by Birmingham Shopping Centre, of which Taylor Woodrow Industrial Estates is a member, on the multi-million pound, 100-shop complex, utilizing space on an eight-acre raft over the New Street Station.

By way of contrast, Taylor Woodrow found themselves at Eton College this year, building the versatile New Hall in concrete which would have astonished the founder, Henry VI. It was designed to be a lecture hall with unusual aids; a theatre for all kinds of school productions; a cinema for both education and entertainment films, and for educational television programmes.

Taylor Woodrow were also much concerned with hospitals at this time. Three were going up already, at Reading, Greenwich, and Armagh; the foundations of a fourth were being laid at Crewe for a 730-bed general hospital for the Manchester Regional Hospital Board; and pre-planning had begun for the new £12½ million Royal Free teaching hospital at Hampstead.

A new company joined the group in 1967, Octavius Atkinson & Sons Ltd. of Harrogate, whose main business was the design and erection of steel buildings.

The new Euston Station was opened by the Queen in October 1968. She arrived at 10.15 a.m. at platform six in the Royal Train from Scotland, just as Queen Victoria had done in the first Royal arrival by train 120 years before. 'My family and I have used Euston Station many times,' she said. 'This opening marks the completion of the most important railway modernization project of this century.'

Taywood Wrightson Ltd., a company jointly owned with Head Wrightson & Co. Ltd., were now appointed by the British Aluminium Company as main contractor, responsible for providing a complete management, engineering, procurement, and construction service for the £37 million aluminium smelter to be built at Invergordon in Scotland.

Among several new orders in power plant construction, the nuclear power consortium, of which Taylor Woodrow is a member, now known as British Nuclear Design & Construction Ltd., was awarded the contract for designing and building Hartlepool power station (1,250 M.W.). In preparation for this, the wire-winding system designed and developed by Taylor Woobrow at Southall to apply 'circumferential prestress' to nuclear and other large concrete pressure vessels was put on to a test rig of the same diameter (eighty-five feet) as the pressure vessels at Hartlepool.

Taking the motorways programme a step farther, a start was made on the £3½ million Aston Expressway contract with the City of Birmingham which provides a seven-lane road link between the city centre and the M.5 motorway. Completions of a new headquarters for Boots in Nottingham and a million-pound central library at Newcastle, and a start on interesting new projects such as a reinforced contract raft over the District Line at South Kensington . . . and into this list of successes came a severe setback.

It occurred due to a very severe gas explosion causing extensive blast damage and partial collapse in a block of flats in the London area. This brought into question the whole justification for industrialized building, at least in regard to high and medium rise buildings.

The Government departments who had pressed the use of industrialized systems and had recommended them, local authorities who had adopted them and contractors who were using them, all awaited anxiously the report of the Griffiths Tribunal which was set up with commendable speed by the Ministry of Housing and Local Government to investigate the disaster.

13(a). Ocean Terminal, Hong Kong, incorporating Asia's largest shopping centre and parking for 2,000 cars.

13(b). Nandi Airport, Fiji

14. Arcon geometry. Port sheds at Doha, Qatar.

15(a). Hotel Inter-Continental, Kabul, Afghanistan. Complete to the knives and forks.

15(b.) Shaft raising at Sizewell. Tunnelling upwards to the sea-bed.

16. Gone is the bang. The Taywood Pilemaster.

The Report of the Tribunal was published in October 1968 and established that the immediate cause of the disaster was a town gas explosion due to the failure of a sub-standard brass nut jointing the flexible connection from the cooker to the gas supply pipe. The Tribunal, notwithstanding its affirmation that 'the building had been put under a microscope' found that 'it must be emphatically stated that no deficiency in either workmanship or supervision contributed to or was in any way responsible for the disaster', and 'taking into account the general satisfactory standards of workmanship, we believe that the whole of the chargehands, the foremen and the Clerk of Works . . . were all effective'.

The Tribunal found that the building was designed to comply with local by-laws and relevant codes of practice but there was no code of practice relating specifically to large concrete panel construction, nor did the Building Regulations and Codes of Practice take into account the possibility of progressive collapse.

The Tribunal recommended *inter alia* that building regulations and codes of practice should be brought up to date and kept up to date. Meanwhile designers of new tall blocks must design buildings to withstand progressive collapse paying particular attention to continuity at joints; they should have special regard to recent research into wind loading; and particular attention should be paid to the effects of fire and structural behaviour on the building as a whole.

Guidance on interim standards to be adopted was given in Circulars 62/68 and 71/68 issued by the Ministry of Housing and Local Government in November and December 1968.

Thus the large volume of industrialized building work being undertaken by Taylor Woodrow-Anglian Ltd, was severely handicapped while intensive full-scale tests were undertaken to establish that the system fully met the new design criteria. Taylor Woodrow-Anglian Ltd., under

G

managing director Geoff Davies, are once more making a significant contribution in the low, medium, and high rise industrialized housing field – such as the £10 million Lisson Green project, the imaginative use by Westminster City Council of the former B.R. Marylebone Goods Yard, to contain nearly 1,500 homes.

Nobody would deny that 1969 was an uneasy one for the contracting industry generally. Yet it was possible for the group chairman to describe it as 'an outstanding landmark in the growth of your company'. The turnover, at £86 million was an all-time record – and this figure covered subsidiary companies only; record profits and a higher dividend; a £23 million irrigation contract in Romania; a new hospital in Sheffield; the first contract in Western Australia for Taylor Woodrow's International Mining Services Division; and the appointment of Taylor Woodrow Property Co. Ltd. as developer for a twenty-five-acre site at St Katharine Docks in the Port of London, described as 'one of the most ambitious private enterprise developments since the Great Fire' – we shall describe it in detail later... all these represented growth.

There were parent board changes during this year. John Fenton, who had been a director since 1935 and, until 1967, deputy chairman, retired. John Ballinger, who had joined Taylor Woodrow in 1946 as a young engineer, became joint deputy chairman of Taylor Woodrow Construction as well as managing director. Sir Patrick Dean, G.C.M.G., former British Ambassador in Washington, joined the parent board as a non-executive director.

A year, you might think, of ups and downs (and one of the downs was a further loss on the Barbican development); and yet one in which 'not unsatisfactory' seemed a more than usually modest understatement. The City, certainly, was enthusiastic. 'It seems characteristic of Taylor Woodrow that a year which has seen small builders folding up left, right and centre, builders' suppliers feeling the draught of both choked-off demand and credit abuse, and even the major contractors in many cases anything but happy, is described by Mr Frank Taylor as "a remarkable one and an outstanding landmark in the growth of the company".'

The writer is a hard-headed investment analyst on the *Investors' Guardian.* 'The TW report,' he went on to say, 'grows more impressive year by year as the global group's stride lengthens on the world construction scene.' He marvelled at a concern so versatile that it included 'such interesting variants of outdoor contracting as augur boring (in this case in the jungles of Indonesia) for nickel sampling. . . . One could go on for pages about TW, becoming almost a legend in the lifetime of the chairman in its contribution to the annals of the British civil engineering industry.'

Noting the chairman's own tribute to the Taylor Wood-row world-wide team and the quality of its people, the *Investors' Guardian* concluded: 'The splendid, and outstandingly consistent, growth record of TW owes much to its success in attracting and keeping some of the most worthy "characters" ever to sport the donkey jacket and safety helmet regalia. Investment in this company is still investment in the best of British contracting and enterprise.'

And so – the seventies. Of new contracts gained in 1970 three stand out as typical of Taylor Woodrow and of the trends they are likely to show during the next decade. Docks, a new nuclear power station, and another major urban renewal project to rival St Katharine Docks in importance.

In driving rain on the exposed flats overlooking the River Forth, the Construction Company's Grangemouth docks project was inaugurated on November 3 1970, when Sir John McWilliam, chairman of the Forth Ports Authority, operating the controls of a Cat. 951, made the first 'dig'. This was on the site of the first of fifty-eight massive concrete monoliths, each weighing 2,500 tons, to form the walls of the new lock which shipping will be able to use in three years' time.

The day after this inauguration, Hammersmith Council approved in principle a scheme for the £50 million comprehensive development of a seventy-seven-acre site in partnership with Taylor Woodrow Property Co. Ltd. Bounded by the north side of Shepherd's Bush Green, the Metropolitan railway line, and the new West Cross route, the area is largely derelict. Here will rise a new 'town' with over 2,700 homes for nearly 10,000 people; a major shopping centre, a 600-bedroom hotel, a 1,000-seat cinema, and new industrial and office buildings. Housing, in units grouped round amenity squares, will rise above industrial and warehousing development, at a height of forty-eight feet, and thus increasing the site to 123 acres of land use. There will be special one-bedroomed dwellings for old people, schools, health centres, and play areas. It is hoped that this 'White City' area will be largely occupied in six years' time, and completed by 1980.

And on December 4 1970, the Central Electricity Generating Board announced that the contract for the new Heysham nuclear power station would be awarded to British Nuclear Design & Construction Ltd., of which Taylor Woodrow is the civil engineering member. Fuelling of the first reactor is scheduled for the first half of 1975 and the completion of the station, which will have a similar output (about 1250 M.W.) to that of the Hartlepool station, is due in 1976.

Power stations remind Taylor Woodrow people of John Ballinger; and amid all these portents of hope and success for the future, they will remember the year of 1970 with sadness. For it began on a note of tragedy. John Ballinger, to whom so much talent and leadership had been given, and of whom so much was expected in a future where it seemed that nothing could go wrong, died suddenly on January 7 at the early age of forty-eight. There was even a strange irony in the date – for January 7 is Frank Taylor's birthday. John Ballinger had seemed to be the heir apparent, and it is generally believed that he would have been elected deputy managing director of the group in June.

A. J. Hill, who had given him his first job 'on two months' probation' in 1946, says of him: 'I shall always remember his grin of delight when we secured the Invergordon contract in which he was largely instrumental. . . . We would all have liked him to have spared himself more, to have cut loose a bit now and again. But it was not to be. It is up to each one of us not to let him down in the fulfilment of what he worked for.'

VIII

The Overseas Story

WE saw, in an earlier chapter, that in the 1930s the scene of Taylor Woodrow's first venture overseas was America, where new ideas of design and layout of housing estates, developed in Britain, were introduced to a country which might otherwise have gone on planning gridded streets and blocks, all in straight lines and numbered instead of named.

For a while after the war it had seemed that East Africa was a probable direction for expansion and export. But no sooner had Taylor Woodrow's advance party arrived in Kenya than there came the famous invitation from the United Africa Group (Unilever) to go into partnership with them in *West* Africa. Nigeria and what we now know as Ghana were in growing need of harbours, houses, shops, factories, banks, schools, universities, breweries.

In Australia, which we left at the end of Chapter V, Taylor Woodrow had begun exploring future possibilities from a base in Sydney. There were critics who thought the group was going to come a cropper there; but perseverance, and Frank Taylor's personal belief in the future there, have long since paid off; and so striking is the Taylor Woodrow story 'down under' that we shall tell it in full in Chapter XIV.

In the Middle and Far East, the group have been increasingly active, ever since the giant development programme started by the Sheikh of Kuwait in the 1950s to raise the living standards of his people. In Europe, there have been projects from Gibraltar to Romania.

The year 1953 saw the formation of Taylor Woodrow (Canada) Ltd. and the buying of a controlling interest in Monarch Mortgage and Investments Ltd. which owns land, blocks of apartments, stores, and houses in the Toronto area, and has its own construction company. This acquisition was a quick, shrewd move by Frank Taylor, who had been disappointed by the small extent of possible British participation in the St Lawrence Seaway scheme, owing to a political decision by the United States Government to go into partnership with Canada. It seemed to make sense to go into real estate instead.

Roy Wykes, an outstandingly able administrator, arrived from Nigeria, where he had been general manager, to become an executive vice-president of the real estate company. With him was Bob Hirst, another veteran of West Africa, who was director-in-charge of Taylor Woodrow of Canada Ltd. Bob Hirst eventually left the group and today Roy Wykes is president of both Taylor Woodrow of Canada and Monarch Investments. Frank Taylor and Les Olorenshaw are on both boards.

Laurie Elion, an accountant at Southall, emigrated with his family to join the new Canadian company and reported back: 'On March 4 1954, our chairman, Frank Taylor, signed our first contract for 120,000 dollars to erect a factory building for the Parker Rust Proof Company on the outskirts of Toronto. A week later we were awarded a 500,000-dollar contract to erect a transit shed for the Toronto Harbour Commission.' From these modest beginnings, all else followed. The construction companies were responsible for one of the world's biggest shopping centres – Yorkdale Toronto–office blocks, university research laboratories, and large-scale mining and hydro-electric developments from Moresby Island in the West to the Mactaquac Dam in New Brunswick.

In West Africa, in collaboration with the Collins Construction Company of Texas, a marine pipeline, taken out to sea off the rocky coast of Accra, had just been completed for the Socony Vacuum Oil Company of New York – a task made difficult by rough surf and cross currents. In April the extensions to Takoradi Harbour, costing £5 million, were opened by the Governor of the Gold Coast. These had involved lengthening the deep-water quay by 1,400 feet to accommodate three more liners, and a new railway system. The most spectacular feature of the whole project was the removal of an eighty-foot high local landmark, 'Cox's Fort Hill'. The spoil from it – two and a quarter million tons of rock and earth – was used to reclaim forty-nine acres from the sea, on which were built a modern railway marshalling yard and cargo platforms.

The art of tendering, and the narrow margins involved, is well illustrated by the preparation of the Takoradi tender. The team working on it argued among themselves for a long time about whether to charge an extra 3d. per cubic yard for earth-moving to cover rising costs. It would make an overall difference of £26,000. Finally Frank Taylor ruled against it. The result was that Taylor Woodrow were able to beat the nearest competing tender by £20,000. 'Had that 3d. been added,' Frank Taylor says today, 'I'm certain we should have lost the contract.'

Operations in East Africa were held up by Mau Mau disturbances: Taylor Woodrow staff, living in very dangerous conditions, had been spending three nights a week patrolling, in addition to trying to do their normal jobs.

Barton Higgs was then managing director of Taylor Woodrow (East Africa) Ltd. He well remembers taking a gun to bed with him, and keeping it in the soap dish while having a bath. Firearms were frequently stolen for sale to Mau Mau, and it was a punishable offence to lose a pistol, which had to be carried at all times. Employing African labour was full of peril; occasionally the body of a loyal African murdered during the night would be found by general foremen on their early morning rounds.

Taylor Woodrow (Canada) Ltd. was not slow to get contracts, and in 1954 was working on a bank, a motel, two factories, and the biggest private project of its kind ever launched in Canada – the construction of three fourteen-storey apartment blocks in Toronto (for Hubert Buildings Ltd.).

In the Near and Far East, Taylor Woodrow Construction was 'building for the future' with new schools and colleges. Typical of their careful preparations was their work for a £2½ million Engineering College for Rangoon University and for a Polytechnic Institute – a complex catering for 1,600 students and described by *The Times* as 'a milestone in the industrial progress of free Asia' – which was carried out between 1954 and 1956.

Extremely high-strength concrete had to be used, and samples of sand from Burma – of totally different texture to the U.K. variety – were incorporated in mixes designed by Taylor Woodrow's research department, and thoroughly tested before the appropriate analysis was decided upon, while climatic conditions experienced in Burma – extreme heat and the cycle of the 'monsoon' season – were simulated on a model of a turtle-shell teak roof.

This roof was for the Engineering College assembly hall and, weighing 140 tons, was regarded as the largest of its kind ever built.

In Kuwait, whose ruler was using oil revenues to create a model state, three modern schools, a research station, and a nursery school were nearing completion, built in partnership

with Abdulla Alireza and Abdul Aziz Saleh. Starting from
scratch, with nearly all materials and plant brought to the
Persian Gulf from England, the team commanded a labour
force of Arab, Iranian, Indian, and Pakistani workers, who
had to learn as they went along, for there were no veterans of
constructional engineering among them. For them two whole
towns of Arcon bungalows were provided. Director in charge
of the Kuwait contract was Francis Morum, and successive
agents were Charles Waggett, Peter Warren, and Ron
Matthews, with Jack Camac (now retired) as office manager.

Taylor Woodrow's sound diplomatic relationship with
Kuwait had been consolidated by Abdullah Alireza's visit
to England early in 1953, and Frank Taylor's return match in
Kuwait shortly afterwards, when he managed to deliver a
speech in Arabic.

In Kenya, Taylor Woodrow's East African Company were
building the new Royal Technical College at Nairobi. But it
was in West Africa that the most spectacular things were
happening. A big water supply scheme in Freetown; law
courts and a bank at Lagos; a power station at Kano. On the
Gold Coast (not yet known as Ghana) Dr Nkrumah, in
February 1954, opened the Achimota–Tema railway, an
eighteen-mile track with three stations, five bridges, and
forty-eight culverts. To build it, an African team, using
modern engineering equipment, shifted 800,000 cubic yards
of earth. This feat gave the lie to those who said 'you can't
train Africans to use machinery, it takes too long'. The
project, linking the future Tema harbour with the main rail
network, trained Africans in the techniques of modern civil
engineering – skills which would be of great value to them in
the future.

Another fifty-mile railway from Achiasi to Kotoku,
through dense forest, bush, and swamp, was also in progress
and, when completed, would clip nearly 200 miles off the

journey from Takoradi to Accra, and open up new farming land for timber and cocoa.

This railway is a good example of what a construction company has to put up with in tropical countries. The figures are daunting enough: up to twenty inches of rainfall a day in the wet season, two million cubic yards of earth to shift, 100,000 cubic yards of crushed stone for ballast, 963 acres of forest to clear. But an eye-witness really brings it home to us: 'The thickly twined undergrowth was cut with machetes, piled and burned; axe gangs sent the great trees toppling, bulldozers dug the roots out and pushed them to the side of the track (this is called "debushing"). In swamp land, winches were used to haul out trees by the roots, and all the time the teams went forward.'

The men lived in Taycot bungalows made of aluminium and lined with hardwood. Two of them were set up at Danso, ten miles along the track from Kotoku, and from that point the men worked back towards the Densu river – one of the biggest obstacles in the railway's path. Some of this Anglo-African team were seasoned veterans who had worked at Takoradi and on the Achimota–Tema railway.

Later, the project was administered from Achiasi, where everyone could live in style: 'Each of the thirty-five bungalows was supplied with electricity, running water and a refrigerator,' said one of the residents, 'and after a hard day's work there was always a shower and an iced drink at the Achiasi Club, with perhaps a film show into the bargain.' Such excellent conditions, which included a resident doctor, were happily reflected in the low rate of sickness, the high level of morale, and the tempo of the track's completion. The tempo, in fact, was dead on schedule.

During Royal tours of Africa, the Federal Law Courts at Lagos were opened by the Queen and the Royal Technical College, Nairobi, was opened by Princess Margaret in

October 1956. In Accra, the Ambassador Hotel was nearly finished. Five storeys high, it was the first really sophisticated hotel ever seen in West Africa, equipped with a bath to each of its 104 bedrooms, eight self-contained suites, and (up to the minute for those days) power points for electric razors. For the first time in West Africa, too, a seventy-foot tower crane was used.

In Australia, Canada, and East Africa, things were difficult at this time; but although the Suez crisis of 1956 had temporarily threatened a setback to Middle East trade, a considerable amount of business had been obtained, particularly in Iran. Taylor Woodrow (Middle East) Ltd. had become Taylor Woodrow (Overseas) Ltd., and had been awarded a contract by the Iranian Oil Exploration and Producing Company for a number of works in the oilfields.

In 1957 the success story was again in West Africa, where the Company was celebrating its tenth year, and above all, in Nigeria, where Tommy Fairclough ('Tommy Drainer') was resident in Lagos as managing director of Taylor Woodrow (Nigeria) Ltd.

Nigeria's most productive oilfield was at Oloibiri, separated from Port Harcourt by sixty-five miles of swamp and forest. Between them a ten-inch pipeline had been constructed for the Shell-B.P. Petroleum Development Co. Because it required specialized 'stovepipe' welding, the teams had to be sent out from Britain – men with curious titles such as stringer-headman, hot-passman, and capper-and-filler, who lived in camps of 'shimbecs' or thatched huts. Theirs was a grim struggle against torrential rain for weeks on end – the worst 'wet season' for years, with flooded rivers; but by using helicopters and timber-surfaced roads they somehow won through.

Work meanwhile was proceeding on the second major harbour contract in West Africa – at Port Harcourt, the chief seaport of Eastern Nigeria, about forty miles inland up the Bonny River. This involved transit sheds, warehouses, new access roads, a reinforced concrete bridge, completely new water supply and drainage systems, the reclamation of six acres of swamp land and seven miles of railway track. Thus, by extending both Takoradi and Port Harcourt, Taylor Woodrow had brought into being two of Africa's major ports. At Port Harcourt they built a modern three-storey office block for their old friends the United Africa Company, as their headquarters.

Several other contracts had also been carried out for U.A.C., notably two trading buildings at Kaduna in Northern Nigeria. At Kaduna too, a £1 million textile factory, designed and constructed by Taylor Woodrow, had been completed for a joint enterprise of the Development Corporation and the Marketing Board of the Northern Region, and David Whitehead & Son Ltd. of the United Kingdom. Its outstanding feature was its steel-framed Arcon structure, supplied by Taylor Woodrow (Building Exports) Ltd., then the largest building ever to be exported.

To complete the Nigerian story, and in spite of the fact that the chairman doesn't drink, Taylor Woodrow were praised for putting first things first when they built their first West African brewery near Lagos. Now, to bring cooling draughts to tropical heat, they built another one at Aba, Eastern Nigeria.

In Fiji, Taylor Woodrow (Overseas) Ltd. extended its empire by starting £2 million worth of extensions to the International Airport; many of the local roads were so soft and narrow that they could not take the weight and size of the company's vehicles, and strange 'Forbidden to Taylor Woodrow' notices appeared everywhere.

But in 1958 the spotlight was back on Africa, this time on French Equatorial Africa, where Taylor Woodrow, in association with the Utah Construction Company of San Francisco and the Compagnie Industrielle de Travaux of Paris, laid plans for the 172-mile Comilog railway which was to cost more than £9 million.

The next two years were particularly difficult overseas. Proposals for a £A6 million office and shop development in Melbourne, Australia, were abandoned as uneconomic. Canada was suffering a general slowing-up due to very severe competition, high unemployment, and over-development. In other overseas countries, Frank Taylor complained in his 1959 Report, 'foreign employers are asking for longer periods of credit at lower and lower interest rates. There is a battle to see who can lend the most money on contracts and carry out work regardless of financial losses'. A new standard form of overseas contract might help, he suggested.

However, Taylor Woodrow had on hand a military hospital in Cyprus, a development for Laurance Rockefeller in the Virgin Islands, and a power station in British Guiana. Business continued healthy in West Africa with the award of a £2 million contract to build the Guma Valley Dam in Sierra Leone. On October 10 1960, Nigeria celebrated her independence with the inauguration by Princess Alexandra of the £4 million wharf extensions at Port Harcourt. These would double the port's handling capacity. Already, while work was in progress, cargoes had reached the million-ton mark in 1959, and as each phase of the work was completed, it was handed over and used immediately.

What do you do if a client suddenly asks you: 'Look, can you cut two months off the building time?' If you are mad enough, as Taylor Woodrow sometimes are, you say 'Yes'. And somehow you do it. Of course, it helps if you have allowed yourself a safety margin in your planning, as I suspect T.W. do. This is what happened in 1961 in Canada, which seemed to be emerging from the slight recession we noted in the previous year. About twenty miles south-west of Toronto, overlooking the north-western reaches of Lake Ontario, is a town called Oakville. This was the site chosen by the Ford Motor Company of Canada for their impressive new central office building.

We have seen, in West Africa, how to drive a contract through on schedule with tropical rainfall as your main obstacle. In Oakville the obstacles could not have been more different: 'General Winter', as the Russians say, with temperatures as low as ten degrees below zero and snowfalls eight feet deep in the severe winter of 1959–60. Excavation had begun in February 1960, and just fourteen months later 600 Ford employees were able to move into the new building, a seven-storey, steel-framed structure adjoining the Ford assembly plant.

The building lies beside the four-lane Queen Elizabeth Highway, and from the visitors' dining-room on the seventh floor, on a fine day, you can see the cloud of spray above Niagara Falls, more than fifty miles away. Forty cases of dynamite were used to blast out the deeper footings for the foundations, and steel erection was finished in seven weeks in spite of the appalling weather – they had to be, because the concrete for the basement walls could not be poured until the steelwork was ready.

Meanwhile, in Freetown, Taylor Woodrow (Sierra Leone) were being inspected by the Duke of Edinburgh, who had come to see how the Guma Dam was progressing. Freetown had always suffered from a water shortage in the dry season (i.e. spring) and a great waste of water in the rainy season. Somehow the water had to be conserved. Incredible as it now seems, Freetown, well-known to every soldier on the round-the-Cape convoys during the war, had water rationing every dry season, which for nearly a century had held back the development of the port and its hinterland.

Then the Guma Valley Water Company, after several investigations and reports, was set up in 1961. By April, Taylor Woodrow had got the contract. 'Within four weeks,' says an observer, 'the build-up of European staff, plant and equipment was in full swing, and the thickly wooded mountainous valley resounded to the roar of machines and shattering crashes as trees were blasted down.' Frank Westaway, at present plant supervisor at the East Lagoon container wharf, Singapore, remembers using four fifty-pound cases of gelignite to shift a tree over 100 feet high and six feet in diameter.

There was a good reason for this haste; for it was known that around June 15 the skies would open and tropical rain would fall night and day for the next three or four months. It did, too; yet workshops and offices were built, and fifteen bungalows to house a staff of thirty people; and bridges were strengthened all along the fifteen miles from Freetown Harbour to bring up the big excavators.

A more than usual amount of rain fell between June and September; yet, somehow, compressors and dumpers were lowered down a steep trail to begin work on the 1,000-feet long tunnel, 16 feet wide, which was to carry the flow of the river while the dam was being built. That tunnel was driven by three teams working twenty-four hours a day, seven days a week, with the swollen river raging outside. High above, two quarries were being established to produce, in three years, 200,000 cubic yards of rock-fill.

Difficult work, and sometimes dangerous; but there were consolations. The site team's bungalows were close to the sea. The Dam Club, a large Arcon building with skittles, snooker, badminton, and a bar, was right on a haven known as Taywood Bay, with its own private beach of silver sand for safe bathing. Many of the team were West Africans trained on other Taylor Woodrow projects in Sierra Leone, Ghana, and Nigeria, where they had gained valuable experience.

17(a). Heron's Hill development, Willowdale, Ontario

17(b). For Export Achievement and Technological Innovation, the first Queen's Award to Industry presented by Field Marshal Earl Alexander of Tunis in 1966. *Left to right:* Harold McCue, chairman Taylor Woodrow (Building Exports) Ltd.; Earl Alexander of Tunis; A. J. Hill, chairman Taylor Woodrow Construction Ltd.; and Frank Taylor.

18(a). Dixon's Blazes. A Taylor Woodrow Industrial Estate.

18(b). Prizegiving 1970. Christine Taylor congratulating Robert Jones, student quantity surveyor, who gained a prize from Liverpool School of Building.

19(a). Cavendish houses, by Taylor Woodrow Homes Ltd. at Kempshott Lane, Basingstoke.

19(b). Industrialized building. Swiftplan at the Garden Centre, Syon Park.

22. Trans-Pennine Pipeline. Pipe-stringing by helicopter helped to overcome the problems of distance and terrain.

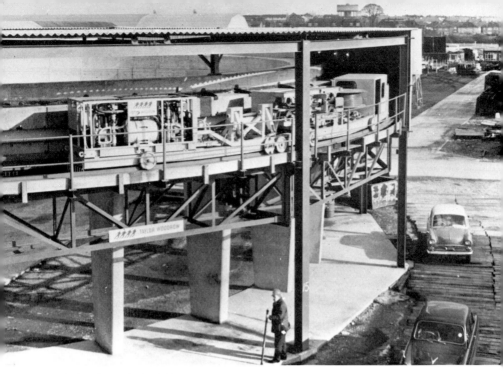

21(a). 'Tayhenge', the full scale test rig at Southall where the wire-winding system for Hartlepool nuclear power station pressure vessels was proved over many months of testing.

21(b). Confrontation at the Barbican. Plain speaking between Frank Taylor and a picket.

We mentioned, a few pages back, the railway to be built by the Anglo-French-American consortium, across Central Africa (known as 'Comilog', because the client's name was Compagnie Minière de L'Ogooué); and the time is ripe to examine its progress. In September 1962, a cargo vessel left the West African Port of Pointe Noire bound for the Ruhr. It carried the first shipment, 15,000 tons of manganese ore, from the mines of Moanda in Equatorial Africa, where there were 200 million tons of the stuff waiting to be mined. This had been made possible by the new railway which, during the past three years, had been driven through and round mountains and dense forests. Six bridges, forty culverts, and forty-five miles of cable railway had been linked across the jungle. The railway was constructed from M'Binda in the Congo Republic to Dolisie, where it connected with the existing Congo-Ocean Railway between Brazzaville and Pointe Noire.

The 3,000-strong international labour force were mainly Congolese, but there were also other African and French, American, British, Belgian, German, Italian, Spanish, and Polish contingents.

Getting the mechanical equipment up from Pointe Noire via the Congo-Ocean Railway, with its steep gradients, sharp curves, and tunnels was difficult enough; to do it in temperatures of around 30°C. with 100 per cent humidity (and in eight or nine different languages) was miraculous. For this contract the senior Taylor Woodrow representative was Peter Hodge, a director of Taylor Woodrow International since 1966, who had joined the group at their West Ham power station contract as assistant engineer in 1948.

Elsewhere overseas, projects were going ahead in Guyana, Trinidad, Colombia, and New Zealand; and in this year, 1962, an important development for the future was the formation, in conjunction with the forty-year-old Bitman

Construction Corporation of New York, of a joint company for building and civil engineering work in the United States.

Charles H. Blitman, chairman of Blitman Construction Corporation (in which Taylor Woodrow hold a forty-nine per cent interest), went into business for himself in 1922 as a general contractor. Since that day, the company's record shows completed projects throughout the eastern United States totalling hundreds of millions of dollars, some seventy-five per cent of which have been on a cost-reimbursement and fee basis. It includes commercial and industrial structures, large housing projects, hotels, shopping centres, roads, water mains, and bridges. In housing, the company is an acknowledged authority on the engineering and construction of reinforced concrete and steel structures.

'Chuck' Blitman, as everyone calls him, shares Frank Taylor's belief in the team approach. He is a licensed professional engineer in the States of New York, Connecticut, North Carolina, Maryland, and Commonwealth of Massachusetts. Graduating as a civil engineer from Rensselaer Polytechnic Institute in 1914, he taught in the Department of Engineering of the University of Minnesota and became City Manager and Engineer of Glasgow, Montana, before enlisting in the U.S. Air Force.

Mr Blitman had been president of the Corporation for many years, until stepping down to the chairmanship in 1968, and was succeeded as president by his son, Howard, also a most experienced operator in this field.

Off the coast of Florida, the Rockefeller holiday resort, in British Virgin Gorda, was going slowly owing to lack of skilled manpower. If the job was to be finished on time, craftsmen must be imported. So Taylor Woodrow went to the north-east of England where there was a shipping depression, and a special recruiting campaign produced sixty carpenters,

masons, and plumbers who, for good money and sunshine, were glad to go the West Indies and complete the project.

And in Colombia, South America, 7,000 feet up in the Andes, work was proceeding, with Bob Aldred as director-in-charge, on the Miraflores earth-filled dam (known locally as 'La Taylor'), hindered by labour trouble and the fact that banditry was the Colombian way of supplementing one's income. Colombian bandits used machine-guns, and it was on the whole wiser to surrender your wallet without a murmur, until such time as you got protection from the Colombian Army. Beset with labour and other difficulties, the contract lost money, but was completed on time and in perfect order.

The doldrums of Taylor Woodrow in Australia were at last broken by a major contract, in 1964, for an ore-loading jetty at Port Hedland, Western Australia. Canada, too, now showed the rewards of perseverance and a buoyant economy, with the four and a half million-dollar Tasu mining project on Queen Charlotte Island, to be carried out in partnership with Albertan and British Columbian construction companies. Another consortium was working on the huge Mactaquac Dam in New Brunswick; and Taylor Woodrow's Canadian construction firm was completing several miles of under-ground railway for Toronto's subway system.

In New Zealand, a minority interest was acquired in a local contracting company, Wilkins and Davies Construction Company, with whom an extension to Auckland International Airport had been built, and the Newmarket viaduct was now under way. This important link in Auckland's multi-million pound motorway system crosses Broadway, one of the busiest roads in New Zealand, and four other streets and a railway line in the central suburban area of the city. It is 2,260 feet long, 70 feet high, and its sixteen spans vary in length from 110 to 200 feet. Its ninety-feet width

accommodates dual three-lane carriageways to carry 3,600 vehicles an hour in each direction.

We have so far said little of the Far East, where factories, offices, shops, and houses were being built in Kuala Lumpur, and an ocean terminal and wharf extensions at Hong Kong, together with extensive earth-moving operations for the Mobil Oil Company.

Hong Kong Ocean Terminal, whose design was co-ordinated by Mike Thomas, Taylor Woodrow International director and chief designer, was opened in March 1966 by Sir David Trench, Governor of the colony. It was quite an occasion, celebrated with Anglo-Oriental panoply. Four junks, decked with flags and banners, were sent to Kowloon Bay to meet the 45,000-ton liner *Canberra*, the new wharf's first customer. As the liner approached, her 2,000 passengers were entertained with a lion dance and firecracker display from the junks, while, on shore, the band of the Queen's Own Buffs played Gilbert and Sullivan.

Commodore Dunkley, master of the *Canberra*, told local pressmen that the Ocean Terminal was 'unique in the world', in that it has a two-floor car park on top. 'It is much finer than other terminals because it has more facilities, including all kinds of shops, restaurants and bars. Hong Kong harbour now ranks with New York, Sydney and San Francisco.'

It had taken nearly three years. The last pile was driven in March 1965, to a crescendo of fire-crackers and the smell of roast suckling-pig, prepared for a celebration banquet on the piling pontoon for the pile-driving crew, who had names like Hiu King Cheung, Chan Yan, and Yee Foo Quock, as cheerful and tough a bunch of men as ever wore the famous 'Four-Men' helmet.

Nine acres of car-parking space; nine acres of air-conditioned concourse and shop space; a drive-in lane on the marine deck for lorries carrying export cargo to drive

straight to loading areas alongside berths; six two-way escalators and five lifts; 'waving galleries' for the public – there was much to be learnt from the Hong Kong Terminal. The buildings, earlier on, had housed the British Engineering Exhibition, part of British Week, opened by Princess Margaret.

Nineteen hundred and sixty-seven was a year of variety overseas, with airport runways in East Pakistan, and international hotels at Kabul, Afghanistan, and Georgetown, Guyana, beginning to rise; flats and houses in Gibraltar; and a new venture in Majorca, where land was being developed by building maisonettes for sale and making available serviced plots for people wishing to build their own homes. New associated companies were formed in the Bahamas for development projects in Nassau and Eleuthera. A port in the Lebanon, a power station in Libya were completed.

'Down under', we detect a definite firming up of progress, as the Port Hedland Wharf is succeeded by a container wharf at Fremantle; and in Wellington, New Zealand, the £3¼ million Thorndon Overbridge also in collaboration with Wilkins and Davies, is taking shape. Aimed at solving Wellington's main traffic problem, this six-lane elevated highway was designed to carry more than 30,000 vehicles a day over the city's northern limits where rail, ship, and road traffic converged and threatened to bring the city's business area to a standstill.

So proud are Wellingtonians of the Overbridge that, as soon as it was open to two lanes of traffic, more than 35,000 pedestrians contributed 3,148 dollars to a crippled children's charity by paying to walk over it for the pleasure of it.

In Australia, after a long haul, the prospects now seemed so promising – especially in Western Australia, where joint companies were set up with the Bond Corporation and Corser Homes. Good – or at least better – news came from

Canada, too, where Monarch Investments, Taylor Wood-row's property development company, reported good results in Ontario. Taylor Woodrow of Canada Ltd., however, were running into difficulties over the Mactaquac Dam, now complete; there were claims which had been referred to arbitration.

West Africa was clouded with the Biafran War. Here, of course, the situation is much improved now; a start has been made on contracts such as the civil engineering side for a new satellite communications link and a new water supply, both near Ibadan. In 1970, a new company, Taylor Woodrow Construction of Nigeria, was formed to concentrate on the mechanical and electrical installations at refineries and similar plants on which activity dated back to the early 1960s.

Back in 1967, the Far East looked brighter, with a £64 million wharf contract at Singapore, in which Taylor Woodrow were to lead a consortium with the Dillingham Corporation of Honolulu, and an interesting mineral investigation project for P.T. International Nickel in Indonesia.

In France, Spain, and Malta interests had been acquired in property development companies, and a new company, Taylor Woodrow (Arts/Loi) S.A., was formed to develop a large office block in the centre of Brussels, with Ted Marsh, a director of Taylor Woodrow Property Co., in charge of development.

Among many works around Heathrow Airport, Myton had built the Ariel, the first completely circular hotel. The group very nearly repeated itself in 1969 when it completed the 109-bedroomed, circular Pegasus Hotel in Guyana: the nine-storey, 120-feet high, circular tower is the highest building in Georgetown. The contract was won by Ken Maxwell's design team on sheer originality. The advantages of a circular hotel are that the hollow 'core' takes lifts, services, drains, etc., there is less distance to walk to rooms, and the rooms them-

selves are wedge-shaped, giving a feeling of light and space.

Project manager was Keith Powell and contracts manager, Tom Norris, Taylor Woodrow International production director for the Caribbean and West African areas.

Many thousands of miles away, at the foothills of the Hindu Kush, a still larger luxury hotel – the first of its kind in Afghanistan – was opened. The 200-bedroom Hotel Inter-Continental, finished thirty-three weeks ahead of schedule, had been built under a complete 'turnkey contract' or package deal by which the contractors 'design, construct, equip, and finance'. Here the project manager was Barry Ing-Simmons – who saw much of both hotels as he moved there from Guyana – with, as contracts manager, Ron Whitehouse, now a Taylor Woodrow International director.

In California, against strong local competition, the Taylor Woodrow Property Company was named as developer designate for the first stage of a major urban renewal project known as the Yerba Buena Centre in San Francisco. The scheme includes a seven-storey office block with ground-floor shops, and a car park for 200 cars.

A significant development overseas in recent years is the penetration by Taylor Woodrow of the Eastern European countries, beginning with a large textile mill in Hungary. Then, on November 2 1965, an aeroplane took off from London to Budapest carrying the first of three batches of 18,000 day-old chicks for a model poultry farm in Hungary. This was a supreme example of the Taylor Woodrow 'package deal' and its advantages to a foreign client.

'We do everything – design, construct, equip, train the staff to operate the factory – the lot,' says Alan Langdon, the Taylor Woodrow (Building Exports) director who set it up. 'We even supplied the chickens, which we got from the Ross group, and guaranteed the number of eggs!' The great advantage to a client of this kind of arrangement is that all the

co-ordination is done for him – as Alan Langdon puts it, 'there's only one bottom to kick'.

'It's the *contract* that's all-important when you're operating abroad like this,' Langdon says. 'You have to define and apportion responsibilities so that they make legal sense in both the client's country and ours.' In his experience, negotiations with an East European country should begin three years before the project is due to start. The actual bargaining ding-dong may take six weeks. 'But once the bargain is struck, we generally get excellent co-operation from them: what they agree to do, they do.'

In Romania, where a contract worth £23 million was awarded in 1969 to the Taylor Woodrow Irrigation Group (inevitably known as 'T.W.I.G.'), a consortium of Taylor Woodrow International Ltd., Sigmund Pulsometer Pumps, G.E.C. Electrical Projects and Vickers, 300 square miles of land in the province of Sadova-Corabia in the Danube Valley are being reclaimed for agricultural use, a remarkable achievement that we shall describe in detail in Chapter XV. Bulldozers, earth-scrapers, excavators, and pumps arrived by sea from Britain in the last months of 1969, and in May 1970, 'Twiggy', an amphibious two-seater hovercraft (by Hover-Air Ltd.) was acquired to ease travel over difficult terrain, especially in the muddy lake Potelu area.

There is mud, too, at Malili, in the Celebes, H.Q. of the Indonesian Nickel Project which has been going on for over two and a half years. At night, heavy rains make wheeled transport useless – only bulldozers and sledges can bring up fuel and stores through slushy jungle. This is one of several big explorations being carried out to ensure future supplies of nickel, that indispensable metal needed by so many industries.

Nickel tends to be in places where men have never been before. To get to Malili, you fly for an hour and a half to

Djakarta; hang about there for a day, and then fly to Makas-sar (George Hazell, who is rather anti-Makassar, can tell you all about the bugs in one of the hotels there). Then four and a half hours over the mountains by landrover to Boné, and ten hours by sea in a forty-two-foot launch to Taylor Wood-row's first camp at Susua.

Until recently, no roads led into the area where Taylor Woodrow are surveying and drilling. Shipping is scarce. A small airstrip for light aircraft has been completed, but most team members are flown in by helicopter, and all supplies have to be brought by chartered ships. The area under exploration is about the size of Ireland: there are four field camps and as many as ten fly camps running simultaneously. They are sampling about 100 square miles at a time. The men get their food by truck, boat, helicopter or packhorse. A radio network links all camps, with a link from Malili to the support office in Makassar, from which international cables can be sent.

You can see why the civil engineering and construction industry, particularly when it goes international, requires a quite special breed of men.

Who is Frank Taylor?

I N fifty years, two semi-detached houses in Blackpool have become over 100 different companies, subsidiaries, consortia, spread around the globe. Housebuilding has now been overshadowed by other skills: Taylor Woodrow now describe themselves as 'the world-wide team of engineers, constructors and developers' – in that order. It is not, by some distance, the biggest group of its kind in the world or even in Britain, but it is in many ways the most interesting, and does the most unusual things in unusual ways.

We took a look at the young Frank Taylor in Chapters II and III – the sandy-haired young hustler with blue eyes whose extraordinary energy and intuition had had that strange knack, observable in all natural leaders, of galvanizing people into action. The time has come to meet him, face to face, as he is today, and examine his ideas on industry, man management, and life.

'No, don't shut the door,' he said, the first time I entered his office at 10 Park Street, Mayfair, headquarters of the Taylor Woodrow group. 'It's always left open, even if there's a draught – symbolic, you know – anyone can come and see me, any time.' His wife Christine (she was his secretary and has been with the firm for thirty years) helps to sift the callers. 'People often come and try out ideas on her when I'm not available,' Frank Taylor says, 'and she'll say, "Well, I think Frank will turn that down", or "I think he'd like that".'

The room is not luxurious. It is dominated by a semi-circular desk (designed by Frank Taylor and made for him by Jock Harris, deputy managing director of Swiftplan) and a large globe of the world, presented by an American admirer, in a midnight-blue recess decorated with Betelgeuse, Sirius, and the stars of Orion – designed and painted, as a surprise for him on returning from a holiday, by Jock Harris, who as an ex-R.A.F. pilot is thoroughly accustomed to star navigation. On the window sill are family photographs and the famous Harry Truman notice, 'The Buck Stops Here'. Another placard says: 'Cheer up, things could be worse – you could have *my* job!' There is a great deal of blue – the colour of hope and optimism: the Taylor Woodrow tie, with the 'Four Men' on it, has a blue background. Frank Taylor was, that particular day, wearing a blue shirt.

He lives in a penthouse, as he says, 'over the shop', except at weekends, when he goes to his 960-acre farm near Guild-ford. A day in the life of Frank Taylor begins with a shower, a sharp, ten-minute walk around the block, the selection of a rose for his buttonhole; and he is at his desk around 7.30 a.m. The staff will not arrive until 8.45 a.m. His day is un-likely to finish before 7 in the evening, and even then there may be a working dinner or a late meeting.

His old friend and colleague, Ted Woolf, now retired, says of him: 'He has the magnetism of all leaders – a quality that makes him seem physically bigger than he is.'

'He's tough,' says John Wilson, Taylor Woodrow's top man in Australia, 'not ruthless,' mind you – some people say he isn't ruthless enough – but he's tough in the sense that he'll do furious battle for an idea or a principle. He has soul as well as brain. I should think the only thing that makes him really angry, that he would tear a sizeable strip off someone for, is disloyalty.'

Disloyalty could mean, in the construction business, which

doesn't often 'poach' men from rivals, holding a pistol at your employer's head and saying: 'Another company's offered me £1,000 more to go to them. If you'll match it, I'll stay.' That is not the way to ask for a rise at Taylor Woodrow.

'F.T. doesn't lose his temper,' his staff say of him. 'But you know when he's upset. There's a fearful tension – he goes absolutely white. Sometimes I think I'd rather be sworn at. He can be hurt, too, if he thinks you've behaved badly or ungratefully. But he just doesn't say it out loud.'

One of the qualities he admires in other industrialists is 'not sparing oneself'. Lord Stokes, for example. It is of course his own quality and he expects it of all his team.' I think it is so difficult to reward financially the people who are prepared to burn themselves up,' he told George Bull of *The Director* a few years ago, 'as by the time one has paid tax, surtax, etc. very little is left. Therefore . . . you must use every legitimate additional means of rewarding them. Share option schemes are good for the Company, for the employees, and for our 10,000 shareholders. They have a tremendous effect.

'We believe not only in monetary rewards, but in decentralization; and one important result of this is that we are able to have more companies, and therefore more directorships.' There are over 100 directors in the whole group. On the parent board there are ten directors, of whom eight have come up from the ranks.

Top construction men are never easy to find. We have seen that they have to be a special breed, dedicated, with a capacity for adventure. They get good quarters abroad, so that they can take their wives and families, but they are expected to come back home at least once a year. 'What I am always looking for,' Frank Taylor says, 'is the person who is able to take responsibility, who is able to influence others and take decisions quickly without getting tired.'

People should help themselves, train each other. Frank Taylor has no great belief in Government industrial training schemes – they are an implied criticism of industry itself. Taylor Woodrow occasionally sends young men on courses to Harvard, Ashridge, and Henley, and the company has its own scheme under, Roy Spiers.

Taylor Woodrow has many men at the top who have made it from the bottom; 'but more and more these days it is important to have academic qualifications backed by practical experience'. Even so, 'A director needs as much commercial ability as technical or scientific knowledge'.

Frank Taylor judges men, as he judges most things, by instinct. (One or two of his colleagues think he relies too much on hunch and emotion: maybe, but it seems to work.) Yet he has been known to deny that he has any special ability to pick people at all, as if the team were an organic thing that automatically rejects the weak member. 'I've picked people who have turned out to be unsatisfactory. But bad people don't stay with us.'

How do you get on the board at Taylor Woodrow? Recommendations are made to him, and he has the final say. Some of his directors think he has the first say as well: 'Young so-and-so is a promising chap. I think he ought to be on the board. Put up a recommendation to me, will you?'

One of Frank Taylor's favourite jokes against himself is about his famous intuition. When Taylor Woodrow installed its computer at Southall, Martin Hebb said to him half in jest: 'That's fine, sir. Now we shan't have to rely so much on your intuition in future, shall we?' He was asking of the computer the one thing a computer cannot give.

Is Frank Taylor a gambler? He describes himself rather as a buccaneer. Even the critical move south in 1930 was only a gamble in that he did it without reserves to fall back on.

Calculated risk is perhaps a more realistic phrase. He has been quoted as saying: 'You must make the majority of decisions rightly, even if it's only fifty-one per cent of the time. Otherwise you'll be out of business. The successful businessman must beware of mistaking stagnation for caution, rashness for enterprise, and gambling for speculation.'

Everyone at Taylor Woodrow is encouraged to be self-critical: 'We can never be satisfied, the most we can hope to be is happily discontented.' Asked what mistakes he thinks he has made, he says: 'Lots, I expect.' He wishes he had gone in for property development sooner. Even though his company financed other companies engaged in property development, he concentrated so intensely on construction that he feels he may have missed opportunities: he should perhaps have invested more in land much earlier, and should have held his development projects longer instead of going for a quick turnover. He regrets not having done so during the first post-war Labour Government which restricted housebuilding and made so little land available. For land always appreciates.

How to avoid making mistakes, and how to profit from mistakes once they have been made, is another Taylor theme. It is probably true that the company has never made the same mistake twice, and it would be difficult for anyone in the organization to make a mistake three times – 'The second time, I get cross,' Frank Taylor says. 'A third time – there's no excuse.'

Credit squeezes can ruin small construction firms and harm bigger ones that don't take a long enough view. In his Annual Reports, Frank Taylor is always asking shareholders to judge the group on its performance over the years, not on an individual year. A really bad contract can ruin a year's profit. Yet, in 1965, Frank Taylor was able to say: 'The credit squeeze won't hurt us this year. And I don't think it will touch us next year. We have £100 million worth of orders in

hand with about thirty per cent of our total turnover being done abroad.

'The nightmare of contractors is not to have enough winners to offset the losers. But we have found that some of our losers have been a cheap way of buying experience. Our £50,000 loss on our first London Airport contract was one of the best losers we have ever had.'

There was also a bad loss at Miraflores, Colombia, where inaccurate information about the sort of ground to be tunnelled cost £500,000. 'That came out of the 1964 profits, which would otherwise have topped £4 million.'

For this kind of reason Taylor Woodrow do not give quarterly reports to shareholders. In the first quarter of 1963 – the unprecedented Arctic winter – the group lost £1 million – enough to send a lot of our shareholders rushing to sell – but we finished the year with record profits'. Why so much cash in hand, a financial journalist asked Frank Taylor in 1965? 'We are wise enough to borrow money when we don't need it. It comes in useful to finance our five-year budget of work.'

What other men have influenced Frank Taylor most? Mostly other industrialists (who all, it seems, read each other's biographies). Henry Ford's *My Life and Work* was a powerful influence: 'I liked his doctrine of "small profits, quick returns"' when I was starting out in business,' Frank Taylor says; 'also his respect for science, and his dislike of waste.' Nuffield, too, of course; Andrew Carnegie; and, in earlier days, Herbert Casson and his *Efficiency Magazine*: Casson was several times invited to give lectures to Taylor Woodrow staff. Frank Taylor's reading contains little fiction: it is biography and autobiography that fascinate him – Macmillan's memoirs, all Churchill's writings, George Hutchinson's book on Ted Heath, Francis Chichester – all of them men who have *done things*.

Among his peers, it is probable that Frank Taylor has been

influenced by Sir Godfrey Mitchell, chairman of Wimpey. Wimpeys are bigger than, and in some ways very different from the Taylor Woodrow group, and Sir Godfrey Mitchell came into the construction business by a very different route: in 1919, with the aid of his Army gratuity, he bought an existing firm that in those days specialized in road repairing. The two men – and the two groups – have always been on friendly terms, despite fierce competition, and there is no 'poaching' between them. True, Len Howard, now a director of Taylor Woodrow Industrial Estates Ltd. and one or two others, are ex-Wimpey men, but the transfers were effected with Sir Godfrey Mitchell's prior consent.

It is not easy to define the effect the two men have on each other. Perhaps Christine Taylor got near to the truth when, one evening after the two men had dined together, she said: 'Frank, whenever you have dinner with Godfrey Mitchell, you come home saying: "Wimpey are doing this, that, and the other; we ought to have thought of them too!"' Later, she discovered, through Len Howard, that Sir Godfrey Mitchell reacted in exactly the same way: 'Whenever he had dinner with F.T., he used to come back full of ideas!'

Christine herself is a constant source of ideas. One evening Frank Taylor was wondering how to raise some capital for a possible new building development. 'How about the Standard Life?' Christine suggested instantly, 'they've had our pension money since 1932 – why not talk to them about letting us use some of it jointly?' The result was wonderful,' Frank Taylor says. 'We formed a joint company, and now we're developing many millions of pounds' worth of property together.'

Anti-waste and efficiency are Taylor themes that lead directly to two stories. Frank Taylor has a dramatic way of proving his first point by taking all the loose change out of his pocket and throwing it on the floor. 'Look, that's money. When you see money lying about, you pick it up, don't you?'

20. Industrialized housing. Over 1,000 homes by Taylor Woodrow-Anglian Ltd. at Broadwater Farm, Haringey.

23. Sizewell nuclear power station nearing completion, seen from the cooling water outfall structure.

24. Research laboratories, Southall. Testing a twelfth scale model of the spherical prestressed concrete pressure vessels, designed for Wylfa nuclear power station.

Well, materials have money value. If you over-order, or leave stuff lying around, or let it deteriorate through neglect, you – are – wasting – money!'

In 1945, Frank Taylor visited West Africa for the first time. He saw concrete being mixed by hand, and carried in buckets by Africans on their heads. 'But you're using too many men!' he remonstrated with the site team. 'This whole job ought to be mechanized!'

The team, who had long experience of Africa, shook their heads. 'It's no good, F.T. The Africans can't be trained to use machinery. They are not physically strong enough. It would be too expensive, anyhow.'

Frank Taylor, not yet satisfied, looked again at the kroo boys unloading cement from boats, two bags at a time, and carrying them on their heads, wading through water to the shore. They were able to do this for twelve hours at a stretch.

'Not *strong* enough?' he stormed. 'Look at them!' Frank Taylor got his way: the job was mechanized.

Later, when Taylor Woodrow played the United Africa Company at soccer, Tom Freakley, director-in-charge, showed Frank Taylor the two teams: 'Are the numbers all right, F.T.?' he grinned. 'You're sure there aren't too many?'

'Right,' said F.T. 'The same as in Britain – only eleven a side!'

We have referred frequently to 'Taylor Woodrowisms', sometimes known as 'Frank's wayside pulpit': extracts from books he has read, sayings of wise men, phrases that have sometimes become private jokes or folklore within the group ('not unsatisfactory' is one of them). Another, from Ralph Waldo Emerson, applies to his own life: 'He makes himself wealthy who makes his wants few.' A third is to be found in the introduction to the *Team Handbook*, which is given to every employee on joining: 'Business is not a machine without a soul.' Many others are scattered through *Taywood*

News, the bi-monthly house journal that circulates through-out the group.

Frank Taylor addresses his shareholders as 'Members' and his staff as 'Dear Friends and Fellow Team Members', both in the Annual Report and in *Taywood News*. Very typical of him was his Christmas message at the end of 1967, a year of more-than-ordinary economic crisis: 'In my own mind, I am sure that the only way out of the morass is for the people of this country to realize that work is the salvation of the human race.' *Laborare*, as one of the makers of the Reformation used to say, *est orare*.

To mention Luther in the same breath as Frank Taylor is not irrelevant. We have seen the strength of the noncon-formist atmosphere in which he was brought up, to which he added his own self-control. Methodism demands regularity of life, good works as well as faith, and the ideal that 'no wrong temper . . . remains within the soul'. I asked him about the legend of his non-smoking, non-drinking, non-swearing. We saw the reasons for the first two in Chapter II. The third is due to the headmaster of his school, all those years ago.

'Ben Whiteley used to say to us: "I want you to *speak* well, and by that I mean, have a decent vocabulary. People who swear do it because they can't express themselves, they simply don't know enough words." I've never forgotten that. And it has an influence, I believe! You know people in the construc-tion industry aren't exactly fastidious about their language. But men on the site never use foul language in my hearing. I've never asked them to stop swearing for my benefit. They just don't do it.'

Frank Taylor has the golden gift of being able to relax at the end of the day. 'Sleep is vital. Worry is useless – just get your night's rest. Joe Greenhalgh taught me that. He was a retired Indian civil servant – he'd been managing railways out there. Well, he came back to England on his pension, and he was superintendent at the Primitive Methodist Sunday School I used to attend in Blackpool.'

Today, Frank Taylor goes to church at Blackheath, a Surrey village near his farm, Long Common: he is particularly fond of the children's service, to which he takes his ten-year-old daughter Sarah. He does not talk much about religion, but the feeling is implicit in what he says and writes and does, and in the sometimes emotional phrases which he uses quite uninhibitedly. When John Fenton, deputy chairman of Taylor Woodrow Ltd., announced his wish to retire at the age of seventy-eight in April 1967, Frank Taylor said of him: 'John Fenton is a fine, Christian gentleman. His life, actions and works are a model for all of us to try to emulate.'

These attitudes of mind are found throughout the group, among its friends too. Frank Taylor likes to tell a story of Owen Aisher (chairman of Marley Tile), who once said to him: 'You know, Frank, your competitors used to laugh at all your talk about the team. But we've learned better now!'

If you give loyalty, you are entitled to expect loyalty. It is a word not much used in business today, in this age of mergers, take-overs, insecurity, hiring and firing, nervous self-seeking. In November 1967, Frank Taylor was interviewed in the B.B.C.'s Overseas Service, and he said this about the employer's responsibility: 'You've got to be fair and firm. And it must be in that order. You must first of all give people a square deal and then you can expect the same in return. There are times when you have to make a decision as to what is best for your team as a whole, even though that might hurt a person who might be a good friend of yours. But it's only fair and proper that whatever is right for the organization as a whole, that has to be done. No business organization is any stronger than the strength of its people. It's the quality of the people that counts.'

Frank Taylor hates sacking people. There is a (possibly

apocryphal) story that he once sent for an executive, chatted amicably, thanked him warmly for all he had done for the company. Not until he was outside the door did the man realize that he had been fired.

Nearly all journalists who interview Frank Taylor praise his 'courteous manner and unruffled calm', and suggest that if he hadn't been a builder he might have been a Monty-type general, a headmaster, or a family doctor (not National Health, of course) with a soothing bedside manner, or, more often, a diplomat. 'In fact,' wrote one, 'diplomat he must largely be, with a controlling rein over so many experts in the diverse fields of finance, engineering, design, architecture, construction, equipment; in addition to the ability to discuss a variety of problems with clients whose interests range from atomic power stations to potato crisps, from prestressed concrete in Port Harcourt to tunnelling in Toronto.'

With almost excessive modesty, Frank Taylor tells the Press that his success is due to 'having been lucky and having a good team of people'. But teams have to be picked and led, and there has to be somebody who says 'try it and see if it works'.

Another outburst of modesty is Frank Taylor's charming (and, I suspect, tongue-in-cheek) way of saying that he delegates so much that 'theoretically, I have nothing whatever to do'. This hardly squares with the fact that the board has asked him to stay on as chairman beyond normal retiring age.

In the next chapter we shall try to describe what it feels like to work for Taylor Woodrow; but, in order to understand Frank Taylor's management methods, we must anticipate a little, and look at the *Team Handbook*, a booklet for 'new boys', in which five basic principles are hammered home right from the start:

1. Be a master of three jobs, your own, the one above and the one below you.
2. Do more than you get paid for.

3. Be loyal to your boss and those working under you.
4. Never withhold credit from those under you; rather pass their names forward to the management.
5. Be team-minded. It is teamwork that wins games, battles and business.

We have spoken of delegation. With delegation goes decentralization. Each subsidiary or associated company takes responsibility for its own operations. However, there has to be central control of policy, finance, group buying (where it is economic), personnel, pension schemes, remuneration and bonus, mechanical machines standardization, and sales throughout the group. This is done by the executive board, which meets at 8.45 a.m. on Tuesdays.

'But apart from this,' Frank Taylor says, 'we try to decentralize to the greatest possible extent. We give people responsibility. We tell them clearly what their task is. A man is given the responsibility for a section, and he is judged on results.'

Budgeting is obviously important, and the Taylor Woodrow system was described in an interview by *Business Administration* (February 1970). 'People are given a one-year budget and a five-year budget, and they can spend within the allocation of their one-year budget without going back to the parent board or the executive board. We find that our one-year budgeting is good, and this system of letting people get on with it works out pretty well in practice. You don't keep pulling a plant out to see if the roots are growing.'

Frank Taylor's fervid political conservatism, and his occasional outbursts against 'Trotskyist wreckers' have sometimes been misunderstood. He did not even embrace Toryism until after the war: 'My father was a Lloyd George Liberal. In 1945 I saw that the Liberals were finished, and that Socialism, while full of fine ideas, was simply unworkable.' If he criticizes the abuses of the Trades Union movement, he does so because he wants the problems of management to be appreciated. He sees industry as a trinity of capital, labour, and management, each useless without the other two. If you put a thousand bricklayers in a field without plans, bricks and mortar, or clear instructions, nothing happens. Management is *difficult*; it is *responsible*; if it fails, everything fails. It must read, study, travel, *give* more than is asked of labour. It therefore deserves, and gets, higher remuneration than labour. But there is nothing to stop labour from studying its job, and eventually getting a managerial post. In Taylor Woodrow today, men who were once bricklayers are pulling down £10,000 a year running vast projects all over the world.

An astonishing document, marked 'Private & Confidential', over 4,000 words long, was circulated to all staff in June 1959. It is Frank Taylor's 'Company Policy' (he calls it a 'homily') and it is brought up to date every few years. Here are some points from it:

The group's first aim is to make profits, and to play a progressive role in British industry, especially in exports: 'PROFIT is a clean word. To make PROFIT is very difficult. If we succeed in making PROFIT, then all of us can be pleased, proud and even happily discontented.'

Profit can only be achieved by giving service: 'We are all of us responsible for maintaining our reputation, bright, untarnished and in a good state of repair. Henry Ford once said: "Give the service and all else will follow".' The main elements of service are quality, delivery on time, selling at the right price, and an effective after-sales service.

There follow the Principles of Management, most of which we have touched on already; science; promotion (from within as far as possible); Organization charts; and of course delegation.

'The difference between delegation and "passing the buck" is a desk organizer. When you delegate, you must give clear

and concise instructions.' There follows a quotation from
Eisenhower's book *The European Crusaders*:

'True delegation . . . is not to be confused with the slovenly
practice of merely ignoring an unpleasant situation in the
hope that someone else will handle it. The men who operate
thus . . . are always quick to blame and punish the poor
subordinate.'

Managers according to 'Company Policy' must be 'big
enough to let . . . (headstrong subordinates) . . . go on and
make mistakes knowing that only in that way will they really
learn and grow'. There are paragraphs on 'The Human
Touch', the avoidance of unwieldy bureaucracy, the importance of training and safety; and then, appraisal of team
members: 'During each year all staff Team members will be
appraised by the completion of individual forms. At this time,
all Team members will have an opportunity of stating their
ambitions and will be told face to face with their immediate
leader what we think of them.'

The document ends: 'THINK! THINK BIG!! Carry on
the good work. Many thanks, F.T.'

This is the Frank Taylor who runs Taylor Woodrow.
Another, very relaxed Frank Taylor is to be found at his
country home at Christmas Pie, Wanborough, Surrey. It is a
split-level, modern ranch bungalow which has been skilfully
extended several times. Here he spends weekends with his
wife Christine and his ten-year-old daughter Sarah, who, he
hopes, will eventually go to Queenswood, the girls' public
school in Hertfordshire of which he is a governor.

Here he keeps horses, grows roses (Virgo, Peace, and
Superstar), fruit, and vegetables; and from the wide picture
windows you look out upon rolling farmland, the flashing
windscreens along the Hog's Back, oak trees, and his herd of
beef Herefords. He swims in his heated pool (conventionally
oblong, not kidney-shaped as in Hollywood) and plays

energetic tennis. His telephone number is ex-directory, but he is not in the least shy of visitors: he gives people two duplicated pages of instructions on 'how to get to our house' from London or anywhere else, with a sketch map attached.

It is the house of a builder – functional, elegant, modestly blending into its setting. As I looked at it, I wondered, not for the first time, how the talents of Frank Taylor would have been employed if he had *not* been a builder.

Let me use the words of his immediate deputy, Les Olorenshaw, who said, 'Frank Taylor could never have been anything else but a builder; not just a builder of power stations, harbours, and all forms of construction projects, but a builder in the broadest possible sense. If Taylor Woodrow were just a matter of making money, he would possibly have retired years ago. Or sold out to another firm. But that wouldn't have given him – or us – job-satisfaction. You can tell a lot about a man by what his rivals say of him. There are men on our staff who, when they broke it to their previous employers that they were going to work for Taylor Woodrow, were told, "Well, we're sorry to lose you, but you couldn't be joining a better man than Frank Taylor."

'To try and sum up Frank Taylor in a sentence: he has a complete grasp of human relations, coupled with an inspired flair for leadership.'

X

Working with the Team

'HUMAN relations': if the reader has not already formed some idea of the inner atmosphere of Taylor Woodrow, the author has already fallen down on his task. How do the principles we have quoted work out in practice, all the way down the line?

Take first the main board. This, at present, consists of ten directors of whom eight (including the chairman) continue to take part in the trading activities of the group, while the remaining two are non-executive with additional interests outside Taylor Woodrow. Board meetings, as indeed all other meetings in the group, are organized to last not more than two hours, and to this end many day-to-day matters are delegated to the next stage, the executive board, which is made up from senior executives within the group as well as main board directors.

In addition to these boards and those of other group companies, an interesting and not-too-usual feature is the management development board, appointed by the parent board. It has twelve members, all young and promising executives in subsidiary and associate companies. Its main purpose is to give them the experience of thinking group-wise and to tap their thoughts for ideas that can be of benefit to the group both in the present and in the future. It initiates its own agenda, and arranges visits to works of its own and other companies in a wide range of industry. This junior board's membership is kept on the move by limiting each man's appointment to eighteen months, and the chairmanship is held in rotation. Secretary to this board (and also to the executive board), is Dennis Carpenter, long serving member who is now office manager at the Park Street premises after having worked in East Africa and Australia.

'We try to make this board representative of our various vertical sections across the group,' Frank Taylor told the interviewer in the February 1970, issue of *Business Administration*. 'It gives the chaps who are on the ground an insight into the horizontal. They meet once a month, they're on their own, and they usually choose to meet on Saturday mornings.

'Where they have authority in the vertical structure, in their own sphere, they implement any decisions they take up to that level forthwith. When the decisions go above that level, they make recommendations. Their recommendations come to me on Monday morning, and on Monday afternoon I answer them.

'If I'm not accepting their recommendations, I go to great lengths to tell them why. I've just turned down one idea, and I had the four chaps who had prepared it out to lunch, and I gave my reasons for not accepting their proposal, and they saw the point.' Another body with powers similar to the Taylor Woodrow executive board is that of the Greenham group which includes John Watson, managing director of Greenham Plant Hire; Peter Barrett, chairman of Greenham Sand and Ballast and Ready-Mixed concrete companies; George Borwell, chairman of Greenham's Tool and Tyre companies; and Dick Cogswell, who is chairman of Taylor Woodrow Plant.

Few good ideas ever get lost at Taylor Woodrow. There are companies whose suggestion boxes are filled with rude anonymous notes. At Taylor Woodrow practical suggestions, big or small, are materially rewarded. For example, the winner of the 'Christmas 1969 Group Suggestion Scheme' was Gareth Jones, of Taylor Woodrow International. While working as an engineer on the Mount Newman Iron Ore Loading Pier at Port Hedland, Western Australia, he submitted a design for a piling cap that could be fitted to piling of either twenty-four inches or forty-two inches diameter. This resulted in a considerable saving of time, labour, and cost and enabled pile driving on the contract to be completed ahead of schedule. He won a cheque for £150. Runners-up were three members of Taylor Woodrow-Anglian, whose joint suggestion was for a new type of clamp to fix bolts only two and a half inches from the junctions of walls and ceilings.

The previous summer, John Bryant, chargehand fitter in the motor workshop at Feltham of Greenham Sand & Ballast Co. Ltd., won £100 for the design and fitting of a new tipper body control mechanism for tilt-cab tipper vehicles. There was also an award of £50 to Gee Garthwaite, chief engineer of Phillips Consultants Ltd. (which is another group member) for 'initiative in preparing a metric handbook' for the use of engineers in conversion from imperial units and for designing reinforced concrete members.

In Singapore, where Taylor Woodrow-Dillingham are building the new container wharf, fifty-three year old Mohammed Jamel Kunjasan, from India, has invented a time- and money-saving method of using electricity to allow an extracting machine, in piling, to be drawn clear more easily. George Todd, contracts manager, presented him with his award when visiting Singapore in July 1970.

As long ago as 1957, we find *Taywood News*, the group magazine, recording awards for such ideas as 'an easier method of climbing into excavators', and 'improved illumination in cycle shed'.

Taywood News, although mainly for staff, is also mailed to shareholders who ask for it. It forms part of the group's basic philosophy that public relations begin at home, that a good system of internal communications is continued on and out into the group's friends among the general public. *Taywood News* has been edited for twenty-six years by Nat Fletcher, head of publicity, who, we have seen, first met Frank Taylor in Blackpool forty-one years ago and little thought he would ever meet him again.

On leaving school in 1936, he took a job as a trainee newspaper reporter. However, in 1940, he joined Taylor Woodrow and when the war ended, helped to start the Taywood Sports Club. The club needed a newsletter to persuade non-members to join; and so, in July 1945, Nat was asked to use his previous newspaper experience, and the first issue of *Taywood News* appeared – a modest typed and duplicated job, as befitted those days of paper rationing, which quickly established itself, was promoted to print and half-tone and now has a world circulation of nearly 10,000. In 1948 Bert Rigg, then director responsible for sales and publicity, offered him the job of group publicity manager in charge of all advertising, public relations, exhibitions, and the like. Assisted in recent years by Eric Sadler, the group press officer, a former Fleet Street industrial correspondent, and a small team, he now runs one of the best publicity units in the business.

To keep everybody in the group informed of developments in all other companies as well as his own, more quickly and comprehensively than is possible in a house magazine that appears every two months, there is a fortnightly bulletin. This contains information such as notes on new contracts all over the world; changes of address; the latest quotation of Taylor Woodrow shares; the progress of ideas for the Group Suggestion Scheme; new appointments and staff engagements. All movements of staff between subsidiaries, or departures overseas, including wives and families, and staff on leave all over the world, are listed; also Employment Opportunities for anyone who wants to transfer to another department or company within the group, and houses for sale on Taylor Woodrow estates. At the top of the first page you invariably find the injunction 'Be Team-Minded!' For matters which can't wait, site and office notice boards receive a supplementary bulletin service; at all times, they have a service of progress photographs for display.

Like many other companies, Taylor Woodrow have an 'old school tie' (with the 'Four Men' on it, of course) which is pretty well known around the world. A Taylor Woodrow executive arriving at New York Airport for the first time did not know how to get into the city, and looked about him in bewilderment. He was approached by a courteous New Yorker who, recognizing the tie said: 'As a humble American engineer, I never thought I would find Taylor Woodrow asking for help!'

Courtesy is an order at Taylor Woodrow. 'Never speak to your people as if they were not even dust on the cogs of the organization,' said Frank Taylor in a directive back in 1952. 'I know of nothing more objectionable than to hear an executive calling, as one might call to a dog, to someone to do this or that. Never be afraid of calling your staff Mr or Mrs.' In 1971, however, it is nearly all Christian names – 'I suppose, because we like each other.' This custom finds its way into the Annual Report, and certainly into the annual general meeting, where the directors' place cards read 'Reg Heasman, Les Olorenshaw, Pat Dean (Sir Patrick, according to *Who's Who*), and so on. Only A. J. Hill is known always as 'A.J.', and only the auditor retains his 'Mr'.

Apart from the tie, there is much in Taylor Woodrow to

remind one of school as one would like it to be. Every year in
October, in a marquee near the Taywood Sports and Social
Club at Southall, there is an annual prizegiving and presenta-
tion of long service awards. This has been going on since
1953. The prizes are given by guests of honour, who may be
anyone from Lord Cole, an 'old friend of the family', to
racing driver Graham Hill, who was invited in 1970. Incidentally, this club is quite a surprise packet; lying in wait behind
the offices, plant yards, and other industrial paraphernalia are
twenty-two acres of football, rugby, and cricket pitches;
tennis, squash, miniature golf, and athletics facilities; indoor
swimming pool; Swiftplan club house; Arcon games rooms,
including the country's first permanent building to house an
indoor tennis court with the B.S. Nygrass floor.

At the Prizegiving, long service members with their
families join representatives from technical colleges, schools,
trade unions, youth employment services, and bankers,
directors, and heads of departments of member companies,
parent board directors, and the Mayor and Mayoress of
Ealing, the 'home' borough. Awards are presented for forty,
thirty, and twenty years' service. Completed indentures are
awarded to apprentices. Roy Spiers, personnel service
manager, presents his annual report on education and training. Student mechanical engineers, quantity surveyors,
building apprentices, typists, and clerks are among the
eighty-odd prizewinners. You do not have to be 'top' – you
have to show 'effort'.

Every year young people from sixteen to eighteen years of
age leave school to start a career in industry, and many of
them join companies in the Taylor Woodrow group. They
can train as apprentices in bricklaying, carpentry, heating and
ventilating, electrical work, and plastering. All apprentices
are encouraged to attend technical colleges and study for
appropriate certificates. Some are selected for special courses

on safety, specialist trade work, and the business of industry. Some may even go to Outward Bound or Yorkshire Dales Adventure Centre courses. The group hopes to get some of its future tradesmen, perhaps even foremen and managers, from these sources.

Throughout the team young men are training as civil engineering technicians and quantity surveyors. Many of its site engineers are recruited from degree courses at British universities; others are recruited straight from school and start their careers as draughtsmen, working their way up to become design engineers.

On the sites and in the Taylor Woodrow Plant Company workshops at Southall, Stafford, and Darlington, young men are learning to service, repair, and overhaul the cranes, excavators, vehicles, concrete mixers, pumps, and compressors for which skilled mechanics are always needed.

Schools get only half-day holidays for special merit, but at Taylor Woodrow it is by no means uncommon for people to earn themselves a free fortnight's holiday – anywhere you like within a range of about 900 miles.

This scheme, founded in 1953, is in recognition of 'loyalty, efficiency and zeal', and there are about twenty-five awards every year.

Everyone who joins Taylor Woodrow is given a copy of the *Team Handbook*. Once you have it, there is no excuse for not knowing anything about the group and the way you are expected to behave. A short history of the group and its main achievements is followed by a list of team rules. There are paragraphs on Ethics, Home Purchase (with the company's advice and financial assistance where possible), and 'The Human Side'.

To anyone who has observed the increasing hire-and-fire callousness of contemporary business, this last section is an eye-opener. These people use words like 'warm' and 'human', and *mean* them. Take, for example, the impulse to hand in one's resignation that nearly everyone has, whenever difficulties arise or tempers are frayed. This is Taylor Woodrow's advice:

'Do a little heart-searching. Remember that the grass in the other fellow's field always *tends* to look greener. Before resigning, ask what your future is; talk it over with your immediate chief, your immediate director, and do not hesitate to talk with the chairman of the company, whose office door is always open. Ask for a short holiday: even a long week-end will help, but once a resignation is definitely handed in, it is always accepted.'

There is also a vital paragraph on Safety. The accident rate in the building industry is regrettably high, and safety means a great deal more than wearing a hard hat. The Safety message is dinned into team members on site, in the office, on the back cover of *Taywood News*; on every notice board.

'Site safety' films and talks are given on contract sites by the group safety department under the direction of Dick Vickery, group safety co-ordinator. They are usually held in the site canteen.

The first film teaches the 'kiss of life' method of artificial respiration, for use after an accident has occurred. But the main burden of the talk is accident prevention – the importance of wearing helmets and other protective clothing, the necessity of good site discipline and the avoidance of risks. There are special risks in each part of the industry – for example, a two-day safety training course, under the auspices of the London Construction Safety Group, for crane-drivers in which they are taught 'acceptable slinging practice'.

In 1956 there were 224 workmen killed and 16,587 injured on construction sites. By 1965 there were 229 killed and 44,235 injured. By 1969 the average figures over the years

25(a). Roof level. Topping out the new Glasgow Stock Exchange.

25(b). Rutgers University. The completed science, humanities, and library buildings at Newark Campus in New Jersey.

26(a). Churchill Square, Brighton. A major redevelopment by Myton.

26(b). Hands under the sea. Breakthrough in the conveyor tunnel built to carry alumina from ore-carrying ships under Holyhead Harbour to the new smelter on the mainland.

27(a). Factory for Kuala Lumpur Glass Manufacturers, Malaysia.

27(b). East Lagoon Wharves, Singapore.

28(a). St Katharine by the Tower. A twenty-five-acre development project in London's dockland.

28(b). New beach for Bournemouth. Taybol's dredger *Transmundum II* depositing sand dredged from off the Needles, Isle of Wight, on to the west beach to make good loss by coastal drift.

were 245 killed and 41,000 injured every year. The second film, *The Choice is Yours*, produced in colour for the Federation of Civil Engineering Contractors, vividly portrays many of the hazards which can arise during site work and shows how, by common sense and efficient accident prevention techniques, these dangers can be avoided.

Each team member receives a copy of *Safety Matters*, a booklet of basic safety guidance, at the time of his engagement. Its refrain is:

'Taylor Woodrow does not expect its team members to take chances or work under hazardous conditions.' Thousands of copies have been distributed, and have been widely circulated to H.M. Factory Inspectorate, members of the London Construction Safety Group, National Federations, and other bodies concerned with accident prevention on sites. With Frank Taylor's personal agreement, other companies in the industry have been allowed to reproduce the contents of the booklet under their own imprints.

The group also has its own insurance company, E. & D. Taylor, whose chairman is Ron Copleston and managing director Frank Pursglove. This subsidiary runs the group pension scheme, its B.U.P.A. scheme, and the group continuous disability scheme.

But its main purpose, as is that of the legal department under David Gilbert, is to cover the intricate requirements of the group's wide-ranging activities. These are two of the many sections which, by the nature of their responsibilities and despite the important contribution they make, are not as highlighted as are some of the more spectacular aspects upon which I have touched.

I have sought to mention many of them if only in passing – accounts, estimating, planning; quantity surveying and plant; sales and development; personnel, industrial relations, training, safety, and buying – the overseas sections of which

K

work hand in hand when the business arises of dispatching
men and materials overseas. Among matters not mentioned
are such widely differing aspects as that the group's design
establishments include separate architects' departments for
United Kingdom construction, for Taylor Woodrow Homes
and for overseas projects, and that the group's pension care
includes the appointment of liaison officers to keep continued
contact with retired members.

The same provisos apply to the human element also.
Senior personnel not mentioned hitherto include two Taylor
Woodrow Construction Ltd. directors whose association
with the team dates back to the 1940s – Norman Baker,
an assistant managing director whose duties include overall
responsibility for the company's planning and estimating; and
Jack Lawson, the commercial director, also on other boards
in this capacity. The remaining Construction Company
director – since 1965 – is Jack Ashton, in charge of the
mechanical and electrical division since 1960 and with the
company since 1956.

Divisional directors of the Construction Company not so
far mentioned include Cyril Bayton, in charge of the plant side
and a director of the Plant Company for over twenty years;
Lou Worsley, quantity surveyor for so many contracts; other
long-service members in Leslie Cann, the company's chief
surveyor since 1965 and Bob Smith, commercial manager in
charge of sales section since 1966, who had joined the group
in 1953 to serve initially in Nigeria; and John Wood Rogers,
a more recent recruit in the company – he joined in 1961 –
but with many years in the industry, chief project engineer
of Invergordon smelter project from its inception.

Another Rogers, Jack, has been very well-known for many
years as head of Southall's personnel department, joined the
group in 1946 and has been a Services director since 1960.
Latterly Ken Price and Wilf Brimley have been performing

valuable services also in the important sphere of good industrial relations. No mention has been made either of such personalities as Stuart ('Jeff') Jeffryes, chairman of Phillips Consultants and Stan Tribe, the group chief accountant.

Back in London after many years in West Africa are both Ken Lambert – based in Lagos from 1946 until returning in 1965 – who is chairman of Taylor Woodrow of Nigeria, and Dick Cogswell, who had been in charge of plant operations in West Africa and is now chairman of Taylor Woodrow Plant Co. and a member of Greenham (Plant Hire) board.

At Western House, the International Company's other long-service members include world-wide commuters in Eric Dilley, director since 1958 and Dick Coppock, director since 1967, who head the sales team and John Teasdale, director since 1963, estimating expert on many major overseas projects. Myton's men include Bill Potts and Reg Twaddle, both in the company before it joined the Taylor Woodrow group, now both directors with Reg in charge of sales and Bill a director also for Taylor Woodrow-Anglian. Another Myton director, Roger Raikes, came to them after joining the group in the fifties and seeing much of West Africa. An Oxford rowing blue, he is one of several team members who have won distinction in one form of sport or another.

Sales for Taylor Woodrow-Anglian are the charge of Ken Thompson, whose thirty years' group service includes many years overseas representation of the Building Exports Company. Arcon sales are now directed by Max Green in the U.K. and Alan Langdon overseas, under long-serving 'Tommy' Thomas who became managing director in 1968.

To complete the sales picture, Taylor Woodrow Industrial Estates promotions are marshalled by Gerald Hopkins and the

Greenham group has a whole host of directors and area managers specializing in this sphere.

Then there are the ladies. While civil engineering is very much a man's world (so far – at least one young woman in the group has passed her O.N.C. in building, a predominantly male subject), there is no doubt that neither Taylor Woodrow nor any other big contractor would have reached their present station without untiring assistance from this source and we can do no better than quote the 1970 team members' letter from Frank Taylor which paid tribute to 'our lady members in offices, on sites, in computer rooms, laboratories, canteen, etc., who make such an attractive and efficient contribution to the modern construction scene'. Within the group, in addition to Christine Taylor and Doris Sullivan, whom we have already mentioned, are such personalities as Daphne Hyde, secretary to 'F.T.' for twenty years of her twenty-three years' service . . . she has been round the world in the course of her duties and once spent three months in America on an exchange basis. Then there are Ruth Philip and Marjorie Evans, secretaries to 'A.J.' and Tom Freakley respectively for many years; Elsie Shepherd and Hazel Derbyshire, secretaries to 'L.O.' and George Hazell, respectively. Edna House retired as private secretary from the Greenham group at Christmas 1970, after thirty years with Ted Woolf.

Kathleen Newstead, housekeeper at Southall – her husband is chauffeur to 'A.J.' – has also completed thirty years service, and another husband and wife team, Les and Ivy Jones, retired in 1970 after almost thirty years each – Les, latterly a storeman with Swiftplan and Ivy, senior restaurant assistant at Southall.

All with more than twenty years' service are, at Southall, Olive Mann, assistant to Jack Rogers; Margaret Brown, invoice clerk; and Stella Smith, comptometer supervisor:

at Western House, Kitty Payne, of personnel, and at Park Street, Yvonne Canter, private secretary.

Elizabeth Stilling, an ex-Fleet Street journalist, has taken part in the publicity department growth since the 1950s and Gwen Chamberlain, who retired early for health reasons in 1970, had attained the position of company secretary of Greenham Sand and Ballast and W. J. Lavender; some, such as Beryl Hennigan, for many years with Bill Mangan's teams in the Midlands and North, work away on sites.

These are only a few of those of whose operations behind the scenes I heard tell, and to all others, members of '25 club' or approaching that landmark, who have not been mentioned, my apologies. One name that kept recurring with affection was that of Dr Barnett Solomons, now in his sixties and medical adviser to the group for many years. At a surgery at Southall each Thursday, he examines senior executives at yearly intervals – and it is astonishing how many of them go on weight-reducing diets and cut down their cigarettes in the weeks just before this annual ordeal.

Should there remain any doubt about whether Taylor Woodrow are 'good people to work for', we will end this chapter with a story of Euston Station. In October 1967 (it happened to be Friday the 13th) Frank Taylor arrived at the terminus at 7.45 to catch a train to Wolverhampton to visit the Midland Links Motorway contract. The new Euston Station was a year away from its opening, and having a few minutes to spare before the train went, he took a quick look around. He found a workman named Pat Flynn polishing a concrete skip so thoroughly that he could almost see his reflection in the metal.

'That looks jolly good,' said the chairman. 'Do you work for Taylor Woodrow?'

'Indeed I do, sir,' said Pat. 'I've been with them for ten years and they're a good firm to work for. They're fair and they pay well, though they work you hard. But I don't mind that: you keep your self-respect, sir. I don't want anything for nothing. I like to work hard and get a good wage.' He looked curiously at Frank Taylor: 'Would you be a shareholder of Taylor Woodrow, sir?'

'Actually, I'm chairman.'

The two team members smiled, and shook hands.

XI

The Scientific Approach

FROM the days of mud hut homes, man has had to study new building materials as he discovered new building techniques. From the start, he sought ways of binding sand and stones into a solid mass which could be formed into different shapes. The Assyrians and Babylonians used clay. The Ancient Egyptians discovered lime and burnt gypsum. The Romans mixed slaked lime with volcanic ash – the nearest approach so far to real concrete – and used it for aqueducts, bridges, and buildings, many of which have survived to this day. Lime remained the chief cementing material until Portland cement was discovered about 1800 A.D.

A century later, and for a number of years thereafter, the construction industry was still largely craft based with load-bearing brickwork or masonry the routine methods of building. Rapid technological advances were made possible by the introduction on a wide scale of, firstly, structural steelwork and then of concrete, and in view of Taylor Woodrow's history of development with this material in recent years, it should perhaps be briefly studied.

Concrete, which has so many potentialities, has been defined as: 'a mixture in which a paste of Portland cement and water binds inert aggregates into a rocklike mass as the paste hardens through chemical reaction of the cement with water.' An aggregate can be sand, gravel, or crushed stone.

Ordinary concrete weighs about 150 pounds per cubic

foot. By varying the mixture, you can get a compressive strength of over 10,000 pounds per square inch. By using heavy aggregates, mixtures weighing over 250 pounds per cubic foot can be made for shielding in nuclear reactors. Yet, if you want it for insulation against heat, you can make it weigh only 30 pounds per cubic foot, so light that it can float on water, and be sawn and nailed as if it were wood.

The natural qualities of concrete are enhanced by the reinforcement and the prestressing techniques. Reinforced concrete contains steel rods or mesh embedded in the concrete. Concrete and steel are compatible because they expand and contract at about the same rate as the temperature varies. Because the strength of concrete in tension is only one-tenth of its compressive strength it needs the tensile strength of steel to complement its great strength in compression.

Prestressed concrete is compressed before being loaded by tightening up internal steel wires. Because of this, it can bend under load without developing cracks. This makes it possible to use very high strength steel having a working stress of seventy tons per square inch or more. The steel in a reinforced concrete beam is restricted to under fifteen tons per square inch to prevent any cracks from becoming too wide, which could allow moisture to enter and cause the steel to rust.

Prestressed concrete is not 'better' than reinforced concrete; it behaves in quite different ways. Each material has its own special advantages; there are jobs which each can do best.

Prestressed concrete permits longer spans for bridges; lighter, more slender, and graceful structures which can carry enormous loads. Although its principle was known in the 1880s, it had to wait until better materials and the pioneering genius of a great French engineer, Eugene Freyssinet, combined to demonstrate its practicality in the late 1920s. In Great Britain its use was unknown before World War II but is now commonplace for bridges, elevated motorways, and many other structures.

Taylor Woodrow's major contributions to prestressed concrete have been the development of the huge spherical pressure vessels at Wylfa – the first of their type and size in the world – and the wire-wound cylindrical vessels first used for the Hartlepool nuclear power station.

The Wylfa vessels, of which more will be said later, are not only the largest pressure vessels of their kind to date, but in view of up-dated reactor developments and the change in design to be seen at subsequent stations, perhaps the largest that ever will be built.

To master material such as this, to keep abreast with all advances and to evolve new application techniques, demands research and development programmes on a grand scale, particularly when, as in the case of Wylfa vessels, there were no adequate existing codes of practice and therefore all available design criteria had to be considered objectively, in depth and detail.

Research, design, and construction can be said to be all part of the process of building. While on the one hand, an internationally operating construction group cannot but have long-standing and satisfactory associations with a highly representative cross-section of architects and consulting engineers, including some of the world's most eminent professional men in their fields, on the other hand it must have its own constant programme of research to stay in business. By the 1950s, Taylor Woodrow, in common with other major civil engineering constructors, had its own design and research facilities. Many of these were concentrated in the Terresearch subsidiary, formed as a separate company within the group to co-ordinate the existing specialist activities of foundation engineering and civil engineering laboratories.

Then, primarily to equip themselves for the nuclear power

programmes, Taylor Woodrow considerably expanded their
research facilities. This expansion has been rapid and comprehensive. Today there is an almost unique team of over 500
design and research staff. The entire design team of 1955
would have filled only a small corner of the present atomic
power design department. This, the 'A.P.D.', was formed
in 1955 under John Ballinger's direction and with Reg
Taylor, now director, as chief engineer. Both were to take
part before very long in the formation of one of the world's
first nuclear power consortia. Reg was one of the original
team of engineers and scientists first based in an asbestos hut
at Whetstone, Leicestershire, where there are now the well-
equipped headquarters of British Nuclear Design & Construction Ltd., which succeeded the initial consortium in 1968. He
also helped to start the expansion of Taylor Woodrow's
integrated design and research effort which followed the
award in 1957 of the contract for the design and construction
of Hinkley Point 'A' nuclear power station. With a planned
output of 500 MW, it was the largest of its kind to date. By
1970 it had generated more electricity than any other nuclear
power station of any type.

In the intervening years, as well as his participation in the
growth of nuclear power activities with the rest of the
consortium and the Taylor Woodrow teams, Reg has guided
the expansion of the structural design department, which is
under Ted Woolnough, formerly assistant chief design
engineer in A.P.D., and the civil engineering design department, which is under Bill Stubbs. Ted joined Taylor
Woodrow in the late 1950s from a firm of consulting engineers
after varied experience including a spell in the steel industry.
Bill joined Taylor Woodrow in 1962 after many years with a
leading firm of consulting engineers.

These two departments have designed many important
works including factories, hospitals, office blocks, and marine

facilities. One of their best known achievements was the new Euston Station, where the new platforms and lines were rebuilt while train services were maintained.

Let us have a look inside one of the constituents of Taylor Woodrow's science based effort, the material, research laboratory. Roger Browne, who runs it, says, 'We have our heads in the clouds but our feet on the ground'.

Dr Browne, a civil engineer from London University, spent four years at the Ministry of Supply on an aerial photography project for high-speed reconnaissance before joining Taylor Woodrow in 1960. A tall, slender enthusiast with a Spanish hidalgo beard, he says: 'The nuclear power programme to us civil engineers is like the space programme to aviation – it's as much of a breakthrough as that!'

To Roger Browne, concrete is a living breathing material. Every variation in any one of several dozen properties is worthy of study, calculation, and experiment. In this way a total understanding is achieved of the material and its behaviour from the time it is mixed until thirty years or more ahead.

In his pursuit of knowledge, Roger Browne corresponds with research men all over the world, many of whom he has first met at international engineering or scientific conferences. His enthusiasm has infected his research team, most of whom have their own special field of knowledge. One of the younger members, Roger Blundell, who joined Taylor Woodrow as an apprentice in 1958, now, at twenty-eight, rates as a world authority on the long term movement (known as 'creep') of concrete under load.

In Taylor Woodrow, research is directed to the provision of the best possible technical and advisory and management teams. The work necessary to provide this service varies enormously in kind and degree of effort.

Some work may be 'way-out' such as cutting concrete by

an inert gas working at 30,000 degrees centigrade; other questions may only require a routine analysis of, for example, an expansion joint material. Again, an inquiry from a site may be answered 'off the cuff', while fundamental work for the nuclear power design team may require years of effort. This will start with an intensive study of current knowledge, followed by the development of a theoretical approach to the unanswered problems, which must then be exhaustively checked, confirmed or modified by experimental work.

Materials laboratory work for the design of the Wylfa nuclear power station included a very large programme of research into the long-term 'creep' of concrete at various temperatures. This work, which has been in progress for nine years, involves the continuous recording and analysis of thousands of readings.

Another stage in connection with the nuclear power programme was the perfection of the wire-winding system to be used, first at the Hartlepool and, later, the Heysham nuclear power station, which followed Wylfa. This required four years' work from 1966 onwards by a team of civil, mechanical, and electrical engineers to ensure that the 4,000-plus miles of high-tensile prestressing wire can be wound on to a pressure vessel quickly, evenly, and at the correct tension to resist the internal pressure. The most spectacular feature was the construction at Southall of a full-scale test rig, i.e. a concrete circle of the same eighty-five feet diameter as the pressure vessels. On its own track, mounted on fifteen feet high columns, a Taylor Woodrow built 'train' has rattled around for many hours proving the system and demonstrating it to United Kingdom and world experts. The significance of this advance in technique will be dealt with in the next chapter.

An earlier development at Southall was the Pilemaster silent pile driver, as the result of a *cri de coeur* from Frank Taylor after complaints about noise at a Taylor Woodrow site in Victoria Street. 'For heaven's sake, can't we invent a silent pile driver? With a vibratory or hydraulic mechanism, it should be possible.'

In driving sheet piling, noise is due to the impact, repeated fifty times a minute, of the pile hammer on the steel driving cap. If the piles could be forced into the ground using jacks, noise would be eliminated. But in order to push the piles downwards, you had to have something to push against and it was not practicable to transport enormous deadweights of what is called 'kentledge' around building sites.

So Reg Taylor spent a weekend in November 1960 on the problem and came up with an idea for using the resistance-to-extraction of partly driven piles to enable other piles to be driven farther down. By alternately pushing and pulling a whole group of piles could be driven.

A rig was set up on what is now the Taywood Sports Field. Theoretically it needed a four-foot trench to get started. But it was Saturday morning, and there was no one to dig trenches. So they tried it without – and it worked.

The story goes that Tom Reeves went along to Euston Station to see a test demonstration. After chatting with the site engineer, Tom, who had his back to the Pilemaster, asked impatiently: 'When the hell are you going to start?'

'Start, Tom?' came the reply. 'We've finished – the piles are driven!' Driving rates can be forty inches a minute!

The Pilemaster which, as we have seen, went on to win a Queen's Award, not only makes its own contribution to the relief of nervous tension but according to the *Stock Exchange Gazette* in 1965 'is commercial in terms of costing compared with the plonk-zunk-plonk hammer pile driver'.

As well as the licensing arrangements in Japan to which I have already referred, an agreement was reached with the Munich based contractors, Leonhard Moll, K.G. in 1970, for its marketing in West Germany, Austria, and Switzerland, the

Pilemaster being one of the many Taylor Woodrow technological achievements which have been made available to other firms at home and overseas.

Over the past few years, Taylor Woodrow Construction have held a concrete pressure vessel consultancy contract from the General Atomics Division of Gulf General Atomics, San Diego, U.S.A. They helped G.A.D. to build up their own expertise, leading to the design and construction of the prestressed concrete pressure vessel at Fort St Vrain, Colorado, and the laboratories have undertaken special investigations into concrete properties, prestressing, and wire-winding systems. The contract includes recommendations on construction procedures and cost estimates.

The more job directed research you do, the less troubleshooting you'll have later, say Taylor Woodrow's research scientists.

Trouble can come in a thousand ways – cracking of concrete may have many different causes, for example shrinkage, thermal movement or settlement of foundations. They in turn can be traced back to other causes, such as faulty mix design, poor quality control of mixing, incorrect use of accelerators, retarders or other additives, incorrect detailing of reinforcement or movement joints. Metals may corrode because of ineffective protection, through contact with dissimilar metals or through attack by the fungicide used to protect timber from dry rot. Prevention involves the constant evaluation of new materials, such as glass reinforced plastics, polymer modified cement, new sealants and adhesives, and a continuous effort to ensure that design offices and sites are kept informed of the latest developments in materials and techniques.

New problems continually demand new materials – for example Rescon, a concrete in which cement is replaced by a synthetic resin, was developed to provide a jointless flooring material for heavy industrial use where resistance to acids and other chemicals is required.

Finally, new materials and techniques must be tested. For example, you can cut concrete by a high velocity water jet, which is probably uneconomic at present. 'Still,' says Roger Browne cheerfully (and you don't quite know whether he is pulling your leg), 'it's been used to cut fish fingers!'

Don Langan, M.Sc., a civil engineer who was with Sir Donald Bailey (of Bailey bridge fame) at Christchurch before he joined Taylor Woodrow, stands, he says, 'half-way between actual design and research'. He is in charge of the *structures* research laboratory (we have already seen the *materials* research laboratory). Don Langan works largely with structural models. His department examines design, sites, all relevant data.

'Our job,' he says, 'is to initiate *slow* change. We have to think a good deal about economy: is a material or design unnecessarily strong for its purpose?' On the wall of his office hangs a photograph of a model pressure vessel bursting under overload test. He also solves design problems – for example, in the stilling basins for the irrigation scheme in Romania: what is the best way of slowing down the rate of flow of water in the pipes supplying the supply tanks? And how should floating lattice breakwaters behave when they are used to reduce the height of waves? Up in North Yorkshire there is a deep mineshaft which is probably going to need a special epoxy resin sealant. . . .

More than most civil engineering groups, Taylor Woodrow devote research to specific projects, such as nuclear vessels, large and small, special structures, the behaviour of television towers under wind stress. At the moment Don Langan is fascinated by the possibility of a sodium-cooled fast reactor, which will need a whole new programme of design and research.

Don Langan also has a model workshop, run by Trevor Thomas, who builds architectural models for use in three-dimensional construction feasibility studies and, in co-operation with the design department, for publicity purposes. Not all structures can be analysed solely by mathematical methods. Not all stress patterns can be predicted accurately, because not enough is known about the behaviour of certain materials under stress. So one often has to make models of the structure the sizes of which, from a modest one hundred and twenty-eighth to full scale, vary according to whatever is most appropriate for the matter in hand, as do the constituent materials, reinforced or prestressed concrete, steel and other metals, acrylic plastics, and epoxy resins. The laboratory has a 600-channel, automatic millivolt recorder, which can be used, via punched tape, in conjunction with a digital computer. At what point is a prestressed concrete pressure vessel overloaded? When and in what manner will it fail?

The test facilities of the structures laboratory include a 900-ton test bed for the study of prestressing under various curvatures, and a 1,800-ton test rig to determine ultimate conditions in a wire-wound hoop prestressing system. Three more examples of work undertaken: model analysis of pre-flexed berthing beams for ships; wind tunnel study for 1,100 feet reinforced concrete television mast; investigation into methods of reducing the driving resistance of steel sheet piles in various soils.

Laboratory engineers are available to visit sites and give practical assessments of the strength and safety of damaged or overstressed structures such as tunnel linings, frameworks, and sheetpile walls.

The staff of both the materials and structures research laboratories are in continuous contact with universities, commercial and Government research establishments, and

29(a). Shell and auger drilling by Terresearch from a tower built on Morecambe
Bay Sands, Lancashire.

29(b). The Invergordon aluminium smelter.

30(a). BBC Regional Headquarters, Birmingham.

30(b). Pilgrim Street, Newcastle. A development spanning a major traffic roundabout.

31. Hartlepool nuclear power station, County Durham. The roof goes on above the 237 feet high slipformed towers to provide covered working conditions for the reactor structures.

32. Head Office, Southall seen from the air during the summer of 1969. The main office buildings can be seen spanning the canal in the foreground with, beyond them, laboratories, plant and vehicle workshops, Swiftplan offices, factory and warehouses, the plant yard, the circular wire-winding test rig, and in the background the Taywood Sports and Social Club.

overseas research organizations. They attend and take part in scientific and technical conferences and courses: for example, eleven Taylor Woodrow engineers presented papers at the Conference on Prestressed Concrete Pressure Vessels in London in March 1967.

A combination of hard experience and this laboratory work has been responsible for many of the technical achievements of Taylor Woodrow over the past eleven years.

Slipforming, tunnel driving, dock and harbour construction, and indeed building of all kind, traditional or system, all have research problems. The slipforming operations at the extensions to Meadowside Granary, Glasgow, where the continuous pour required three months of round-the-clock work, were among the largest ever carried out. During the building of Sizewell nuclear power station, Taylor Woodrow introduced to the United Kingdom for the first time a method of raising tunnel shafts, by thrusting upwards through the sea-bed – a method which is being repeated at the Hartlepool power station, and also for a sewage outfall scheme at Lowestoft. Scientific, yes; and yet not so far removed, in spirit, from Frank Taylor's intuition forty years before, when he bought Grange Park Estate, knowing that sewage could be pumped *upwards*.

Taylor Woodrow's chemical and research engineers, technicians, and assistants, may, as Roger Browne says, have their heads in the nuclear clouds, but the majority of problems that come their way are more mundane. Can we use these doors in Gibraltar? What caused this paint failure? How good is this concrete water-proofer? How do these cracks affect the strength of the structure?

When computers came to Taylor Woodrow, structural analysis problems were taken to them; in particular, stress analysis programmes based on the 'dynamic relaxation' method originated by Taylor Woodrow's computer section.

'We encourage our people to take Ph.D.s,' Reg Taylor has said; in the context of computers he means such young men as Alan Welch, who has specialized in this complex subject, and is one of many Taylor Woodrow scientists who have taken part in international conferences.

Although standard structural design programmes are in everyday use by the design departments, owing to the advanced nature of most of its work, the computer section, under Dale Carlton, often uses its own programmes. Programmes required include those for soil mechanics problems, such as the settlement of raft foundations or earth slip calculations; and a good deal of statistical work is produced not only for design and research but also for management information. For example, there is a comprehensive 'concrete quality control sheet' covering every Taylor Woodrow construction site, which enables an unfavourable trend to be identified and corrected in good time. Computers are also used in estimating (such as the optimum size of components) and in critical path scheduling for each site. The only thing they can't foresee (yet!) is the weather, and many a suffering project manager has asked them to do their sums all over again.

To the non-technical layman, one of the weirdest sights is the 'digital plotter' which executes design drawings, or maps the contours of a new site, all from computerized information, and more quickly and accurately than any draughtsman can.

Another Southall based, design set-up is that of T.W.C.'s flourishing mechanical and electrical division, under director Jack Ashton. His people see to the complete and detailed design, and supervise the construction and installation, of what you might call the 'nuts and bolts' side of the business. Some of their projects at home and overseas form part of larger group contracts; others are placed direct by outside clients. One of three main streams of activity is mechanical installations – process plants, chemical works, pipelines, and heating and ventilating services generally. The others are electrical installations and instrumentation and control. The needs of the oil, chemical, power, motor, and food processing industries are among those which have been keeping them busy.

The Construction company's estimating and planning department (director-in-charge Norman Baker) can appropriately be mentioned here also, as its engineers and estimators, in addition to the preparation of tenders, carry out research into new construction methods and materials, development and design of projects. Most of their work is concerned with plant and method, their cost and effect on design. Backing it up are such tools as fundamental methods research. A 'Job Planning' service helps sites with particular problems that might arise.

While much of scientific effort is Southall based, Taylor Woodrow International have, at their Western House head office, a large design department which has been responsible for an enormous volume of overseas work. Some examples, in addition to the £23 million Romanian irrigation project include international hotels in Afghanistan and Guyana; a wide range of industrial projects and marine works in many parts of the world. More recently too, they have undertaken survey and exploration works for the mining industries in Asia and Australia.

Also at Western House is the sixty-strong design team of Phillips Consultants. This firm was acquired by Taylor Woodrow in 1958 and has been responsible for the design of numerous buildings at home and overseas, ranging from that Afghanistan hotel to potato crisps factories in England and Scotland.

Myton and Taylor Woodrow-Anglian have their own design responsibilities, while Taylor Woodrow (Arcon) Ltd.

carry out at their Welbeck Street, London office (where Don
Kershaw is chief designer) designs on which are based many
commercial and amenity buildings for Africa, Asia, and other
overseas countries.

Taylor Woodrow put a lot of research into the S.A.F.E.G.E.
Monorail. This and the Pilemaster attracted unusual attention
five years ago. Taylor Woodrow still hold exclusive licences
in the United Kingdom, Canada, and various parts of the
Commonwealth for S.A.F.E.G.E., and Frank Taylor saw this
as a key tool in helping to solve the world's growing problems
of transport congestion. 'If you think of it as being in the
same stage of commercial development as the hovercraft,
that is a very good parallel,' he told a financial journalist at
that time.

There might have been a sixteen-mile route from the centre
of Manchester to Ringway Airport: the Ministry of Transport
certainly seemed interested. There might have been another
from Birmingham city centre to Elmdon Airport, and
another from West London Terminal to Heathrow. Many
were the aldermen who visited the test track at Chateauneuf-
sur-Loire. But it was not to be: the M4 cut the London–
Heathrow road journey to half an hour, and the Piccadilly
Line extension was begun in April 1971.

In the final analysis, the monorail seems to have been held
temporarily in abeyance by sociological forces which Taylor
Woodrow, with their concern for environment, well understand – the still-further, above-ground disturbance and
complication of people's lives, and a new kind of noise. It is
better to have loved and lost . . . and technical enthusiasm is
never wasted. It may happen yet, either in the United
Kingdom or in a younger and less cluttered land, and Taylor
Woodrow maintain their faith in its feasibility. Meanwhile
'It was a useful exercise', Taylor Woodrow people tell you;
and they generally add: 'the publicity wasn't at all bad'.

XII

The Power Game

A NUCLEAR reactor is essentially an enormously powerful source of heat, which is generated at a carefully controlled rate. In the type of nuclear power station used in Great Britain for Wylfa and all earlier stations from Calder Hall onwards, the heat is transferred from the reactor by carbon dioxide gas, which is circulated by powerful blowers through heat exchangers. A heat exchanger is like the boiler of a conventional power station but uses hot gas heated by the reactor instead of from burning oil or coal. It transfers heat from the hot gas to water, converting it into steam. This drives a turbine, which turns an alternator; the alternator generates electricity.

A nuclear reactor sounds as simple as boiling a kettle the way Frank Gibb explains it. Now managing director of Taylor Woodrow Construction, Frank joined the Company in 1948 and has been responsible for many power stations, tunnel, and motorway contracts.

A boiler: that is really all it is. The art lies in achieving smooth co-operation between hundreds of civil, mechanical, and electrical engineering designers and the integration of research, design, and construction in the whole complex effort, which results in a new power station producing its designed output on time.

How did Taylor Woodrow get into the power game? In one sense, they have been in it since the early years of the war, beginning with opencast coal contracts. Their first thermal power station contract was at Warrington 1943. Just as it had once been difficult to get a War Office contract to build Army camps ('But you're housebuilders – you've no experience of camps!'), so it was not easy to get on the tender list for power station work.

Among the jobs at Warrington was to add five feet diameter cooling water pipes and other extensions to an existing power station: the pipes had to be run under an administrative annexe without disturbing the people working there! Such works as railway sidings, hoppers, and coal handling plant followed.

'It was the war that gave us our first big expansion,' A. J. Hill, joint deputy group chairman, remembers. 'After it was over, we thought the boom was over. We had to find new ways of carrying on, new special skills to offer.'

'A.J.' had experience of power stations with another construction company before he joined Taylor Woodrow in 1938. He and the Australian-born Geoff St Barbe Connor (whose untimely death in 1952 robbed the group of a skilful administrator) went together after power station contracts. Alexander Gibb, the consulting engineers, who were active in this field, looked at them perhaps rather distrustfully: the Warrington job, they felt, wasn't real power station work. Nevertheless, a new contract was awarded in 1946 for £90,000 worth of power station work at West Ham and soon this grew into a £1·4 million worth which went on until 1954 and included a concrete wharf to the River Lea, turbo-alternator blocks, with superstructure, pumphouse, and rail works.

'The interesting thing about West Ham,' says Tom Reeves, who was sub-agent, 'is that so many of the team on it have made good since. As well as Frank Gibb, there is Dick Puttick (now a parent board director); Len Brooks, deputy chief surveyor; Ron Matthews (now a director of T.W.C. Midlands); Peter Hodge and Sandy Cheyne, engineers.' The last two are now both directors of Taylor Woodrow International and Sandy is also a divisional director of Taylor Woodrow Construction.

And of course Tom Reeves himself, who had come a long way since the Mulberry Harbour caissons. So confident of his power station expertise was he that when the prospect of a new station at East Yelland, Devon, came up, he cheerfully telephoned A. J. Hill: 'If you don't give me that job, mate, I'll come round and knock your block off!'

The block is still there: Tom Reeves got the job and, like the heart and soul production wizard he is, carried it out magnificently; and today, the man who at fourteen sold newspapers to save enough to buy his first tool kit, has been a director of Taylor Woodrow Construction since 1953.

A big (over £4 million) Midlands contract at Castle Donington was among those which followed; then North-fleet in the south, High Marnham in the Midlands in the same year – 1962; later, West Burton – the Midlands again, one of Europe's first 2,000 MW stations – and many others. At one time, Taylor Woodrow were working on thirteen power stations at once, and by 1970 they had contributed to the building of nearly two dozen thermal, hydro-electric, and gas turbine stations, mainly for the Central Electricity Generating Board, but including the complete renewal of Lots Road, one of London Underground's suppliers, while keeping it operational.

Not that there weren't occasional setbacks, such as the fire which one winter's night in 1957 destroyed reams of carefully prepared paperwork – including much of the Hinkley Point 'A' design – and the temporary building in which they were housed, in the middle of the Southall premises. This building had been the hurriedly established headquarters of the new

Atomic Power Department, but within a few hours – by noon – teams of draughtsmen were 'billeted' in other offices, anxious inquiries were assuaged and the A.P.D., as we have seen, went on from strength to strength.

Earlier, the winning of the contract for the world's first full-scale nuclear power station at Calder Hall, Cumberland, had elements of drama that would translate easily to the television screen. The cast consisted of Frank Taylor, A. J. Hill, Tom Reeves, and John Ballinger, a young engineer of twenty-nine who had already achieved the rank of agent; Sir Christopher Hinton of the U.K.A.E.A., and Sir Charles Mole of the Ministry of Works. Sir Charles wrote a 'job specification' for the man he wished to see in charge of the contract (the main requirement was that he should be 'very senior') and demanded to see a list of the entire team who would be working there, complete with biographies and photographs.

The Taylor Woodrow argument was that since nuclear power stations were new, and nobody had any experience of them, it was pointless to have a 'very senior' man in charge. This was a job for a young man, free from preconceptions, backed by an experienced team; in particular for a young man of proven brilliance who was destined for the top.

On seeing John Ballinger's photograph, Sir Charles exploded: 'I don't want a baby in charge of my work!' The initial meetings were between Frank Taylor, A. J. Hill, Sir Christopher, and Sir Charles, with the rest of the team waiting in another room.

'That's the man I want!' said Sir Charles, pointing at Tom Reeves. But Frank Taylor risked all by insisting that the contract would be turned down altogether if his recommended team and its leader were not accepted. After all, Tom Reeves would be supervising and supporting. . . .

A series of toughish arguments, some verbal, some written,

followed between Tom Reeves and Sir Charles before Sir Charles accepted John Ballinger. His doubts about youth quickly turned into admiration of the tall, fairish, very serious, young engineer whose professionalism he could not fault.

The only observer who had any criticism to make was the Soviet leader Mr Malenkov, who visited Castle Donington and Calder Hall in his March 1956 tour of British power stations. 'Why does assembly work take so long?' asked that cheerful brick-dropper. 'We will send you some special literature on that point.' Banished subsequently to a Russian power project in Kazakhstan, he may even have learnt something himself.

Calder Hall, which was for the United Kingdom Atomic Energy Authority, led to the consortium principle for building nuclear power stations for the Central Electricity Generating Board.

As we read in Chapter VI, Taylor Woodrow's civil engineering works there began in 1953 – pioneer operations they were, too, for this was the world's first nuclear power station to turn out electricity for peaceful purposes. Later the English Electric and Babcock and Wilcox companies, also widely experienced and internationally operating, were awarded Calder Hall contracts in their specialist fields. Then experts from the three companies opened their historic sessions at Whetstone, made their plans and assessments, initiated their design programme, which can rightly be described as epoch-making in its field, and witnessed in 1955 the formation of one of the world's first consortia equipped to carry out the complete central design, planning, construction, and commissioning of nuclear power stations. This consortium was the original English Electric, Babcock and Wilcox, Taylor Woodrow Atomic Power Construction Co. Ltd. It is now known as British Nuclear Design & Construction Ltd., and is one of the two U.K. 'survivors' of a number of consortia once in being. The three original members have remained close colleagues throughout, and the headquarters is still in an enlarged establishment at Whetstone, where its departments of physics and laboratories carry out a wide range of development and proof testing. Whetstone research is backed by that of its member companies.

The last name change was made in 1968, in the light of a detailed Government survey of United Kingdom industry, resulting in the addition of the U.K.A.E.A. and the Industrial Reorganization Corporation as members, the latter on a temporary basis.

The consortium tendered for Bradwell and two other stations but were unsuccessful. Perhaps this was just as well, for what they eventually achieved in 1957, was the contract for Hinkley Point 'A', whose 500 MW planned output was appreciably larger than those for the Bradwell and other predecessors.

This was the consortium's first design-and-construct nuclear power contract. Taylor Woodrow were now fully involved in the world of large-scale nuclear power, where a tender may cost £500,000 or more to produce and require a large van to carry all the documents. This was also the signal for the rapid build-up of the atomic power design department described in the last chapter. From now on Taylor Woodrow were carrying full design responsibility for millions of pounds' worth of civil engineering work.

Hinkley is a windswept headland, nine miles from Bridgwater, Somerset. There two reactor buildings were to rise 180 feet, and other principal buildings included a huge 740 foot long turbine house.

Among the purely civil engineering works which began in 1957, the 160-acre site (of which 40 acres were occupied by this station), was protected by a 3,500 feet long sea wall which helped to reclaim eleven acres of land, and the little hamlet of Combwich, four miles along the coast, was transformed into a new port by the building in six months of a new wharf, to enable heavy equipment to be shipped from the north and Scotland.

The next stage after the design for Hinkley Point 'A' was the award in 1961 for the 580 MW nuclear power station at Sizewell, Suffolk, not far from Lowestoft. This station was again the most powerful to date, but it was also the most compact so far designed, with both reactors housed this time at either end of a composite building 193 feet high. Then came Wylfa, on the northern tip of the Isle of Anglesey, a much larger proposition.

Planned in 1962 and begun in 1964, Wylfa was to have been built by different consortia in two parts, each with a capacity of 500 MW. But there were doubts about the wisdom of this: would not a single station, with 1,000 MW capacity be more economical? Taylor Woodrow showed that by 'optimizing' the size of the pressure vessel such a two-in-one station could yield 1,180 MW. This enabled the consortium to quote a very competitive price for what undoubtedly was a still further major advance in nuclear power station capacity and technique.

The designed output is 590 MW from each of the two reactors, twice that of any other 'Magnox' reactor.

In the short space of nuclear power station activity, by this time, many design innovations and technical improvements had already been successfully introduced, all making their contribution to more efficiency, more economy.

At the time Wylfa was being planned, there came a transformation in the approach to pressure vessel construction. Each new pressure vessel seems the last word until someone (usually, it seems to me, Taylor Woodrow) designs a better one. It is almost a case (to quote Lord Mountbatten on defence

weapons) of: 'If it works, it's obsolete.' Each new reactor is a
step forward in a constant effort to make pressure vessels
smaller, more efficient, safer and cheaper, easier to build,
easier to inspect and maintain.

At Hinkley and Sizewell, steel vessels were used. They
were made of three inch and four inch thick steel plate
respectively and were so large that they had to be transported
to site in sections. On site they had to be welded together,
section by section, minutely inspected by X-ray for welding
faults and surrounded by a massive concrete biological shield
to support them as they were erected. These vessels were
straining welding techniques of very large vessels to the limit
and Wylfa, if built in steel, would have required five inch
thick plates. Then, in 1960, the French built the first experimental prestressed concrete vessel. Two years later Taylor
Woodrow were planning Wylfa, with the world's first
spherical prestressed concrete pressure vessels.

As we saw in the last chapter, these are the largest monolithic prestressed concrete structures yet built (ninety-six feet
internal diameter) and may remain so. They are each constructed from 30,000 cubic yards of concrete of exceptional
strength and quality. The minimum thickness of the walls is
eleven feet. The external shape was painstakingly evolved to
accommodate the many plant penetrations and prestressing
components. The prestress operation occupied almost the
whole of 1968. The 1,338 tendons required for each vessel
were manufactured in a site fabrication shop from a total of
over 3,000 miles of specially prepared high tensile steel strand.
Each tendon consisted of a total of 36 0·6-inch diameter
strands; each strand in turn a seven-wire composite.

When working at its operational pressure of 385 pounds per
square inch, the internal force to be maintained by one of the
concrete vessels at Wylfa is over 180,000 tons.

More advances were to come, again including improved

pressure vessel techniques. By 1968, cylindrical vessels were being designed for the Hartlepool station. Here, of course, the circumferential stressing is achieved by winding multiple layers of pre-tensioned steel wire around the outer periphery. Within the next few years, Hartlepool and Heysham will be in commission with nuclear furnaces contained in ninety-six feet high concrete cylinders, each bound with over 4,000 miles of wire like giant cotton reels. Why change from a sphere to a cylinder, from golf balls to cotton reels?

'We consider it a leap forward,' says Dick England, now divisional director of Taylor Woodrow Construction, who went as chief engineer to the Hinkley station, and was contract manager at Wylfa until 1967. 'The disadvantage of a spherical vessel is that you can't remove the boilers.'

An advantage of the Hartlepool-Heysham prestress system is that it enables the standardized-type boilers to be installed within the walls of pressure vessels, after having been fabricated and tested under factory conditions rather than on site, and that they can be removed for examination and maintenance, a facility not so far available for other gas cooled reactors.

Other new construction methods initiated at Hartlepool and to be followed by Heysham, include the early and rapid erection of six 237 feet high, slipformed concrete towers to form the main structural elements of the reactor hall which is quickly roofed over to give weatherproof protection for the installation of equipment. The plant layout places the joint station operating and control services closer both to the turbine hall and fuel handling area than had hitherto been conceived.

We saw, in the last chapter, how many engineering problems can be solved in the laboratory before construction of a nuclear power station ever begins; how, in particular, entirely new techniques, of which wire-winding of pressure vessels is one of the latest, had to be developed because industry could not supply them. Programming of thousands of different activities over perhaps a five year period, in all weather conditions, has to be planned by computerized critical path methods, plus high organizational skill, in order to get men, plant, and materials together at the right time.

Inside the reactor itself, a diverse mixture of trades is competing for space – specialists in core erection, biological shield erection, boiler construction, pipework, circulator assembly, insulation, and so on – nearly all with their difficult problems of access and a constant risk of getting in each other's way, unless there is tight control.

Calder Hall, Hinkley, Sizewell, Wylfa, Hartlepool, and now Heysham; all sited near the sea, because the cooling system for the turbine condenser needs an enormous volume of water to be pumped up from the sea and returned to it. (As it returns it warms the sea, which attracts shoals of fish, to the delight of local anglers.)

Cooling water systems present special, often unique, problems. Hinkley Point 'A' was designed for thirty-five million gallons of water per hour. It was decided to build, in an on-shore dry dock, a reinforced concrete island ninety-six feet in diameter, weighing about 3,000 tons. This was floated to a point in the Bristol Channel one-third of a mile off-shore and lowered on the sea-bed on eight 'spud leg' supports. This structure connects with intake tunnels about 2,000 feet long, and the system was designed to serve a second station which would need fifty million gallons an hour. C. S. Allott and Son were consulting engineers for the cooling water works.

Dangerously rough seas often hindered work at Hinkley. In October 1957, seven men were marooned on a sixty-feet high drilling tower in the Bristol Channel during a storm while boring into the sea-bed for rock samples. They were

John Hardy, John Flynn, Bill Gates, Denis Birch, John Gielty, James Sullivan, and John Rawsthorne. They saw the waves rise to within eight feet of the top of the tower, the fourteen-foot dinghy in which they had come from the shore was waterlogged and the motor damaged. They hoisted one of their red jackets and lit diesel oil in a bucket as distress signals. They rigged up a tarpaulin for shelter, and hung desperately on to the scaffolding. Fortunately they were spotted by someone who alerted the Weston-Super-Mare lifeboat, which took two hours to reach them.

To meet the problems of cooling water each nuclear power station usually has its own different solution. At Sizewell, which uses twenty-seven million gallons of water an hour, vertical intake and outfall shafts were constructed by driving them from underwater tunnels upwards through the sea-bed by hydraulic jacks.

While at Calder Hall, 300 feet high cooling towers were built, some C.E.G.B. stations with which Taylor Woodrow have been concerned have required the driving of tunnels in compressed air. This operation, the need for which is not, of course, confined to nuclear power station construction, is one of the more difficult jobs in civil engineering. To keep the water out, air pressure has to be balanced with the water pressure. If you come out of the tunnel too quickly into normal atmospheric pressure, you get a peculiar complaint known as 'the bends'. Entry and exit therefore have to be slow and controlled by an airlock.

Among the men who specialize in 'soft-earth tunnelling' are a special race of cockney-Irish miners, earning £90–£100 a week, many of whom live in south-east London and frequent the Elephant and Castle public house, which is where you have to go when you want them.

Apart from these and other specialists, power station teams are versatile, and there is a hard core of Taylor Woodrow men

- crane drivers, carpenters, concreters - who can turn their skills to many other tasks.

Sometimes the remoteness of the site raises an unusual logistic detail. For the United Kingdom Atomic Energy Authority's 240 MW prototype fast reactor at Dounreay, Caithness, where, as at Calder Hall, Taylor Woodrow is carrying out the civil engineering works, the customary studies to ensure the smooth flow of essential materials and equipment included coping with the carrying of cement from the Thames to the port of Scrabster 'round the corner of' Pentland Firth but this hardly rated as a problem. From the civil engineering point of view, particularly interesting features on the thirty-five acre site were the foundation and the cladding works. After excavation to sixty-feet depths, in deep rock and to true profile, construction of foundations included the drilling of sixty-five rock anchors to depths of forty feet into the old red sandstone underlying the concrete. These units were stressed to 130 tons load, and in addition forty prestressed anchorages were used within the concrete liner of the reactor vault.

A steel framed, 615 feet long building rising to up to 120 feet high and of irregular shape containing the power house complex is the main structure on the site. This was clad with concrete units, some of which weighed up to twelve tons. These were manufactured in a temporary nearby factory, installed on an old wartime aerodrome.

The casting beds - six large steel vibrating tables each weighing four tons - were set on twenty-five-pound rail tracks accurately lined and bedded on the old runway surface.

Some 350 panels manufactured under controlled conditions were generally twenty feet long and their design, together with infill concrete, was such as to ensure air-tightness of the active areas and the provision of bracing at roof level to ensure the final stability of the structure.

5. Commercial Union (*right*) and P & O headquarters buildings,
London, examples of 'umbrella' construction.

6. (a) H.V.I. Luboil Plant, Stanlow.

6. (b) Cargo Tunnel at Heathrow Airport.

From the viewpoint of group profitability, power station work has given Taylor Woodrow the stability of major, long term contracts. This applies not only to the nuclear programme but to the more traditional coal- or oil-fired stations, which we must not forget: contracts for these are continuing.

In addition, as has been already indicated by Tom Reeves, power station work has been one of the sources providing a training ground extraordinary for men later to make their mark in the group.

Some we have already met, others we have still to meet, such as Frank Carr, deputy chairman of Taylor Woodrow Construction (Northern) Ltd. first seen at some of the Trent Valley (High Marnham, West Burton) power stations. Frank's managing director at Darlington is Ian Mickler, one of the wave of young men who joined at Hinkley Point; another was Ken Williams, who moved towards the end of 1970 from Invergordon to be supervising contracts manager at the new construction base at Airdrie.

Earlier on the scene had been Sam Davies, now a director of Taylor Woodrow Construction (Northern) who actually joined the group in 1938 and, after war service, was at the Yelland power station from 1949–51. Ken Sales, now project manager at Hartlepool, was also on power stations from soon after the war; Dick England, who joined at Nottingham power station in 1953; and Alan Grimmett, at Castle Donington in 1955, one of whose most recent projects has been the Midland Links Motorway. John Russell, who was with power stations from 1954, wound up at Wylfa before his transfer to Southall in 1970 as personnel manager; and Bill Jenkins, project manager for the whole of the Company's work at High Marnham power station, was appointed in 1971 construction manager for Taylor Woodrow's Mechanical and Electrical Division.

Among the advance party which left London for Calder

M

Hall on August Bank Holiday Monday 1953, was Vic Blondell, works manager who succeeded John Ballinger as agent in 1955, went on to start Hinkley Point and Sizewell (and was finally, as a change from nuclear matters, at Euston before he retired in 1969). Also of Calder Hall 'vintage', Eddie Varcoe was chief engineer: he is now deputy chairman, Taylor Woodrow (Midlands). One of his assistant engineers was Noel Lakin, who after becoming chief design engineer, A.P.D., at Southall, was appointed a divisional director in 1971.

It has not been uninterrupted progress, of course. The sudden death in 1962, when he was project manager for Euston, of Tom Smith, at the early age of forty, was a keenly felt loss. Among his contributions to Taylor Woodrow, he had been in charge of off-shore works at Hinkley Point 'A', and subsequently became contracts manager to supervise all the civil engineering contracts under Tom Reeves.

Detailed attention to a power station complex continues at all stages. We will take as an example the proof pressure test, conducted in December 1968 at Wylfa under the control of Dr D. F. Seymour, chief mechanical engineer of British Nuclear Design & Construction Ltd. A combined team of engineers and technical staff from the three member companies spent four days literally 'underground'. They were on duty continuously during the main pressure test, sleeping and eating in specially built accommodation at the test control centre in the basement of the reactor hall. There were observers from the C.E.G.B., Associated Engineering Insurers, the Ministry of Power, and several supplying companies.

Special gauges were used, capable of detecting movements to an accuracy of one part in a million. Movements and temperature of concrete and steel were minutely recorded. At each stage of pressure raising, all this data was processed, graphed, and evaluated, involving some 2,000 computational operations. The verdict: good for the next thirty years at least.

We have written much about Wylfa already, and in Chapter XI we saw how the Southall laboratories came into being, and owe their progress to the fact that nowhere else in industry could the necessary know-how be bought. Those prestressed concrete pressure vessels could never have been designed or tested on existing information about the behaviour of concrete. You can predict it mathematically, but you cannot be certain of its safety under new conditions without rigorous testing.

The future seems clear. The cost per kilowatt of power generated has for the reactors of the Calder Hall type, known as Magnox after the magnesium alloy used to 'can' the natural uranium fuel. Important advances in design have been made.

Wylfa was the last of B.N.D.C.'s Magnox stations, Hartlepool is the first of the advanced gas cooled reactor ('A.G.R.') type. The A.G.R. is a progressive development of the Magnox. Together, these types have formed a secure and reliable foundation for Britain's expanding nuclear power programme. While work is going on behind the scenes to ensure that the A.G.R. will continue for some time to come to meet the needs of electricity authorities, progression to the high temperature reactor, for instance, would seem the next logical step. B.N.D.C.'s capacity also extends to such projects as the Steam Generating heavy water reactor and the sodium cooled fast reactor.

The consortium is already responsible in the United Kingdom for 4,760 Megawatts of 'nuclear power for peaceful purposes'. It will be strange indeed if it does not find itself, in the next twenty years, building more such power stations, of various outputs, in many other countries.

XIII

'Paddy's Wigwam'

'I SEE the Cathedral every day. You can loathe it or you can love it, but you can't ignore it. That is the measure of its impact.' Thus Bishop Harris, Auxiliary Bishop of Liverpool, in the presence of the architect Sir Frederick Gibberd, at a press conference in June 1966.

Once you have actually stood inside it, you love it. The Liverpool Irish, half affectionately and (because it represents a breakthrough after many years of intolerance) half bitterly, have nicknamed it 'Paddy's Wigwam'. It is the Roman Catholic Metropolitan Cathedral of Christ the King, the first Cathedral in concrete and, strangely, such is the oneness of things, it could only have been built by a construction company that knew a good deal about nuclear power stations. From Calder Hall onwards, Taylor Woodrow had become thoroughly accustomed to dealing with concrete in unusual shapes.

'We found ourselves working in a new geometrical field,' says John Wilson, project manager on the first (constructional) phase, and now director of Taylor Woodrow International in Western Australia. 'Our nuclear power experience was important in several ways. Quality of concrete, for instance. For nuclear power projects we had produced special concrete that could withstand radioactivity and very high temperature. It also has to withstand smog: Liverpool has one of the most corrosive atmospheres in the United Kingdom. Nuclear power stations had also given us a lot of experience of dry-expansion joints.

'Behind the whole Cathedral project was what I would call a special "philosophy of design" – Sir Frederick Gibberd's design has an almost Nonconformist simplicity. (Yes, I *am* a Roman Catholic, but the fact that I was project manager was purely coincidental – I just happened to be available at the time.)' He also happened, A. J. Hill points out, to have the essential skill and experience.

After two years, John Wilson was succeeded by Peter Stewart (to enable John to go to Australia), who carried the job through to completion, assisted by David Taylor and Derek Shepherd. Throughout the construction period the general foreman was Tom McNulty, and the chief engineer was Bill Baines (now with the northern company). Tom Reeves was director-in-charge.

John Wilson found time, on his last home leave from Australia, to visit the Cathedral again. 'The acoustics are superb,' he said. 'I got the organist to play for me – that low D, more of a vibration than a note, and the swells – you should hear them!'

Did he have any special emotional feeling about building a cathedral? He smiled: 'No, civil engineers don't think that way. It's the same if you are building a hospital – you are not thinking about all the sick people who are going to be treated there. You are just obsessed with the desire to make a first-class job of it.'

In 1850 Liverpool got its first Roman Catholic Bishop, the Rt Rev George Hilary Brown, whose co-adjutor and eventual successor, Dr Alexander Goss, commissioned a design for a cathedral from Edward Welby Pugin. It was Gothic and dominated by a spire, rather like Salisbury Cathedral. Sir John Betjeman would have loved it. But funds were short, and the project was abandoned after the

Lady Chapel had been built. It is now known as the Church of Our Lady Immaculate, in St Domingo Road, Everton.

In 1911 Dr Whiteside became Liverpool's first Archbishop, and his successor, Archbishop Keating, revived the idea of building a cathedral as a memorial to Dr Whiteside. But it was *his* successor, Archbishop Downey, who commissioned a design from Sir Edwin Lutyens, and coined the slogan 'A Cathedral in our Time'. In 1930 a site was purchased on Brownlow Hill and four years later the crypt was begun. It was near completion by 1940, when the war stopped all further work.

Lutyens' Cathedral was designed to seat 10,000 people; to cost £3 million, and to take twenty years to build. Archbishop Downey is described by Patrick O'Donovan of *The Observer*, author of *A Cathedral for Our Time*, as 'a diminutive, top-hatted portly man who became one of the magnates of the city'. He had big ideas. So did Sir Edwin, who provided for 'the largest dome in the world over a vast Cathedral in the style of the English renaissance. The Archbishop and the architect were prepared to wait 400 years for its completion.'

Sir Christopher Wren was one of the very few cathedral architects who have seen their work completed in their lifetime. Hard-headed Taylor Woodrow men say that this was because he offered a 'turnkey' contract – design, construction, and finance all in one package deal. The designer of St Paul's was, above all else, a practical man and an administrator.

After Lutyens' death in 1944, Adrian Gilbert Scott was appointed to carry on the original design. But by 1952 the estimated cost had risen to £27 million. Scott produced a new design, based on the crypt and costed at only £4 million, but nobody liked it much.

In 1957 Archbishop (now Cardinal) Heenan decided that the era of stop-go-stop cathedrals must end. He wanted an open competition, and he was absolutely tough about the cost: it must not exceed £1 million. It must be (and here he adapted Downey's phrase) not only '*in* our time' but 'a Cathedral *for* our time' built from materials *of* our time. Competitors were told: 'The priests and the people of the archdiocese of Liverpool will beg God to enlighten you.'

The story is told in Sir Frederick Gibberd's *Metropolitan Cathedral of Christ the King, Liverpool*. The new Cathedral was to seat 2,000 and the high altar was to be the focus of the whole design: 298 architects from many countries submitted schemes anonymously, and No. 253, Sir Frederick's, was the winner. After eighteen months of planning, building began in October 1962. Four years later the structural work was largely complete, and the solemn opening ceremony and consecration was held on Pentecost Sunday, May 14 1967.

It is clear, from Sir Frederick's account, that he is what Taylor Woodrow would call 'team minded'. Having designed several power stations himself, he knew what they had in common with his Cathedral, and the kind of technicians he would need to work with. The army of architects, engineers, quantity surveyors, helmeted construction men, artists, manufacturers, and suppliers are mentioned in great detail.

He has at home a photograph of Scott, architect of the Albert Memorial, in morning coat and top hat, inspecting that edifice with the builder and foreman respectfully agreeing with everything he says. 'The scene is ludicrous compared with a site meeting today, where it would not be unusual to find thirty highly qualified specialists hammering out the construction problems as a team.'

One of the most important decisions taken by the consulting engineers, the quantity surveyors, and the architect was to call in the contractor at a very early stage. 'We looked,' says Sir Frederick, 'for a contractor with exceptional organizing skill and technical expertise.' There was no question of tendering. Sir Frederick preferred to deal with a sub-committee. Three contractors were short-listed.

'We eventually recommended that Taylor Woodrow Construction Ltd. should be appointed. I had, metaphorically speaking, a hot line with A. J. Hill and Tom Reeves.

'Occasional informal discussions quickly developed into daily consultations at all levels between those engaged on the job, and we were fortunate indeed that John Wilson, who was to become project manager on site and his second in command, J. H. "Bill" Baines, were both available for some months before the work actually started.'

The cathedral is a vast cone of 350 feet diameter at base, sweeping to a central tower whose topmost pinnacles are 300 feet above the ground. A giant tower crane, protruding through the top of the cone, was a familiar sight in Liverpool during construction.

'I seized the opportunity of giving Liverpool two crowns,' says Sir Frederick. His own cathedral and Gilbert Scott's earlier Anglican cathedral now dominate the city's skyline.

His own, with its sixteen pinnacles, symbolic of a crown of thorns, has a colour scheme of white and pale grey to give it a clear outline and to contrast with the dark sandstone of the 394-feet tower of the Anglican Cathedral. The two buildings are about 800 yards apart on high ground in the centre of the city.

The Metropolitan Cathedral has been conceived with sixteen reinforced concrete trusses (or flying buttresses) forming a circular nave with the high altar in the centre: the congregation is grouped round on three sides, and the choir on the fourth. The trusses support a tapering tower filled with coloured glass. Eight small chapels are built between the trusses and separated from them by stained-glass windows.

Among many gifts from business and industry, Taylor Woodrow contributed the Archbishop's Throne. Their

Christmas card for 1966 showed a photograph of four priests cheerily greeting four men in Taylor Woodrow helmets.

The stained glass in the 'lantern', designed and executed by John Piper and Patrick Reyntiens, has 156 panels which make up a complete colour spectrum: the three main points of light, symbolizing the Trinity, are yellow from the North, blue from the South-East and red from the South-West.

Because this is a cathedral 'for our time', there is provision for a tea-room, lavatories, car parking, and two television galleries.

Despite all the decorative work by artists of our time, it remains simple. It is not an art gallery. 'It is as modern as the newest ship in the Mersey,' says Patrick O'Donovan, 'and as old as the darkest underground chapel in the Levant. . . . It is an act of ordered worship.'

XIV

Australian Venture

MANY British ventures have foundered in Australia because the initiators hadn't done their homework. There is the story of a bath manufacturer who spent £2 million on a factory in the wrong state before he found out that many Australians prefer showers, and there are vast areas where there isn't enough (or any) water.

Taylor Woodrow were once accused of 'ignorance of local conditions' by Oliver Marriott in his book *The Property Boom*. 'The Western Market project in Melbourne . . . is discussed in hushed tones by those respectful of their rivals' mistakes, or with glee by the brash. The scheme . . . never got off the ground.'

Taylor Woodrow are the last people to claim that they never make mistakes. What they do usually claim is that they seldom make the same mistake twice. Ever since 1951 they have felt (and it wasn't only Frank Taylor's famous intuition) that something big was going to happen in Australia – on one side of the sub-continent if not on the other.

They began in 1951 from a base in Sydney, and they began with housing. Two years later it seemed that a New South Wales Government contract for 5,000 houses was in the bag, and a team went out to form Taylor Woodrow Australia Ltd. with John Hanson in charge. This time it was bad luck, certainly no lack of homework; they were suddenly hit by a credit freeze imposed by the Australian Government, and the project had to be abandoned. However, a thirty-six-mile pipeline had been built for Shell in 1952, and other industrial projects such as the lecture theatre and associated extensions for Sydney's Nuclear Science Institute.

'If at first you don't succeed, try once again, and then try something else.' This old business precept is at least consoling when you are trying to make a go of things in a new territory. So Taylor Woodrow went and had a look at Western Australia; and in 1964 there came a big breakthrough when Taylor Woodrow International started work on the Port Hedland contract. They had not only discovered how completely different Western Australia was from Victoria and New South Wales: their arrival there happened to coincide with a mining boom. They were no longer a cocky pommy company competing with native Australian builders, but civil engineers doing what, to them, with great experience of harbour works, came naturally.

By 1965 they had established an office in Perth. By 1966 the ore-loading jetty at Port Hedland was complete.

The jetty, situated off-shore from Finucane Island, forms the western boundary of the harbour. The island is the ore shipment assembly area for the Mount Goldsworthy haematite mine which is about seventy miles away. First shipments of the iron ore were made during May 1966, only fifteen months after the scheme left the drawing board, and the first 25,000 tons went to Japan.

Construction at the jetty was carried out for Goldsworthy Mining Pty. Ltd., with Utah Construction and Mining Company (of San Francisco) as main contractors. Taylor Woodrow International Ltd. were responsible for marine installations, and dredging was done by Utah Dredging Division. Before the dredging, the harbour had been little more than a tidal creek, and due to its configuration tidal conditions and currents were exceptionally severe. Between

August 1965 and February 1966, 400 piles had been driven – with difficulty, not only because of the tides, but because of dense hardpan and limestone close to the surface of the sea-bed.

Piles were driven with the aid of a 90-foot by 60-foot pontoon and a special universal pile frame designed by Taylor Woodrow International. This rig is capable of driving piles sideways and fore and aft, up to 120 feet in length.

In this inhospitable tropical region, says Bob Aldred, 'the site team started off living in caravans. When we eventually sent John Wilson to live in a bungalow in Perth to look for new business, the lads envied him his luxury.'

Life is now a little more comfortable at Port Hedland, as long as you stay on dry land. However, the Australian team have recently been working in the ocean approaches where sharks and venomous sea snakes abound, and the temperature reaches 110 degrees Fahrenheit in the shade. In 1970 they completed a contract for the design and construction of the navigation aids marking the entrance to the harbour. The contract was let by Mount Newman Mining Co. Pty. Ltd. and administered on their behalf by Bechtel Pacific Corporation Ltd.

The navigation lights enable 100,000-ton iron-ore bulk carriers to be brought into the harbour by day or night and at all states of the tide. The farthest beacon is eight miles off-shore and rises 100 feet from the sea-bed. Weather and sea conditions off the north-west coast of Australia are so hostile that work is sometimes impossible for three days out of seven. Yet Taylor Woodrow's marine civil engineering team some-how carried out intricate piling and assembly operations day and night, in shifts, finishing the construction work in only seven months.

Back to John Wilson in his £5-a-week bungalow in Perth, where he stayed on with a small nucleus of technicians after the main Port Hedland harbour had been finished in 1966.

Why no shopwindow in Perth? 'People aren't impressed by plushy offices when they don't know your work' John Wilson says. It is not so easy for pommies to gain acceptance down-under. Taylor Woodrow were already convinced, as Bob Aldred says, that 'development follows mining', and there were obvious outlets for Taylor Woodrow skills and experience of drilling, soil sampling, irrigation. Mines need ports, ports need railways, along railways townships tend to spring up. . . .

The team had done an excellent job at Port Hedland, but Australia was not yet beating paths to John Wilson's door. John Wilson, forty-three, six feet four inches, who makes energetic gestures as he talks, has spent most of his professional life abroad, starting with B.P. in Iran and moving on to Pakistan before he joined Taylor Woodrow Construction in 1958. Such a man is not easily discouraged. He was 'tendering like mad' for further work – and at last got the contract for a new container wharf at Fremantle. 'This was our turning point in Western Australia,' he says. 'Our contract was on a design-and-construct basis, and we did the whole job in thirty-four weeks – six months ahead of schedule. Speed of that order had rarely been seen before in Australia.'

The container berth, 800 feet long and with a minimum depth of 36 feet of water alongside, consists of a steel pile structure supporting a heavy prestressed concrete deck. The deck can carry twenty-five ton containers stacked in threes, and the berth is equipped with a forty-five ton capacity gantry crane and other facilities for rapid handling.

Another contract soon followed at Fremantle. In order to make maximum use of the container terminal, the Port Authority urgently needed a parking berth for waiting ships. The solution was to build 'dolphins' – V-shaped jigs projecting from the shore and supported on single temporary piles, with fenders against which waiting vessels can lie.

Fremantle was followed by the very difficult Carnarvon Jetty job, 600 miles north of Perth at Cape Cuvier. It was needed for another mining outlet – to ship salt to Japan. Few people who have not seen it have any idea how wild the coast is here. 'It's famous for shipwrecks,' John Wilson says, 'and has been so for centuries. When we arrived there, we found it hadn't been surveyed since 1810, when it was charted by a French sea captain called de Freyssinet (no relation to the concrete man). There was no village there, only our camp. No telephone – we kept in touch with Perth with our own radio. Nothing was on our side – we even had to blast an approach road 200 feet deep to get at the jetty at all.'

Taylor Woodrow's re-entry into the rest of Australia was cautious. It is probably true to say that Frank Taylor himself felt wary about it, certainly as regards general construction. But he was listening hard. George Dyter and Colin Hunt, of Taylor Woodrow Property, flew out in 1966, and reported possibilities.

In 1968, Frank Taylor and his wife made a business visit to Australia, where Christine has relatives, with whom she hoped to have a few days' holiday which never materialized. F.T. looked at Fremantle and Perth, and on the spot decided to go ahead in Western Australia. Taylor Woodrow were to buy land and property. The way to work in Australia was to link up with Australian companies. He flew on to Sydney, and there took other far-reaching decisions.

The result was the formation of a number of new companies in association with prominent Australian development companies – Taylor Woodrow Bond Pty. Ltd. (with Bond Corporation), Taylor Woodrow Corser Pty. Ltd. (with Corser Corporation), and Taywood Hughes Pty. Ltd. (with Hughes Bros. Pty), whose interests are in New South Wales.

Back home in England, Frank Taylor said to Nat Fletcher, his publicity chief: 'Nat, they don't know enough about us down there!'

Now it was Nat's turn to see for himself. He went with Bob Aldred, and the result was a chain of receptions all across Australia, with British High Commissions in each state being extremely helpful. It was a real 'whistle-stop tour', taking in Perth, Adelaide, Melbourne, Canberra, and Sydney. There were meetings with Government Ministers and officials, industrial and business executives, and developers, telling them about Taylor Woodrow's work throughout the world. Films were shown, power stations were discussed, likewise docks, refineries, roads, factories, offices, and housing – all the things the expanding Australian economy will be needing.

'Australia is a country by turn harsh and sweet, from the outback areas of the north to the verdant plains of the south,' Nat reported. 'The immense natural resources have to date been scarcely scratched. The potential is enormous: iron ore, oil and gas, potash, bauxite, timber, gold, nickel, and land development.'

In Perth, Taylor Woodrow Bond have built International House, an eighteen-storey office block containing comprehensive amenities including a motel-type hotel, and another wholly-owned development company, Taylor Woodrow has built City Arcade, an extensive shopping and amenity complex in the centre of the city.

New homes are going up at sites such as the Port Kembla area of the magnificent Lake Heights estate on the Illawarra coast of New South Wales, and across the continent at the equally spectacular Lake Joondalup, Western Australia. Land is being bought for the development of shops, offices, flats, and houses; a whole new beachside township is planned for Yanchep, near Perth. Specialists from Britain, Canada, and

other areas have been supplementing the nucleus Taylor
Woodrow team in Australia.

The former Western Australian Premier, Sir David Brand,
says, 'the keys to success are work, capital and the spirit of
the modern pioneer. If you have these qualities, you will
be welcome – and needed.'

Brian Trafford, now a director of Taylor Woodrow Property Co. Ltd., married to F.T.'s eldest daughter by his first
marriage (another son-in-law is Simon Kimmins, who is in
the international export and banking business) has also
recorded his impressions of Western Australia. From Carnarvon he drove seventy-five miles over rough country, only
two and a half miles of which had a tarmac surface. He and
John Wilson had radioed the site of the Texada Mines potash
contract, which they were to visit, 'to notify our expected
time of arrival, after which a search party would be sent
out'.

Cars in this district are usually fitted with 'roo-bars', a sort
of extra fender to cope with wandering kangaroos. Often the
track is obscured by sand drifts.

When they eventually got to the contract, they went
straight into a site meeting 'which went on until about one in
the morning'. One of the items on the agenda was 'Action
to be taken on receipt of a 30-hour cyclone warning'.
Accommodation was in air-conditioned caravans, tethered to
steel stakes and perched on the edge of a cliff. Reveille was at
0515 hours (a klaxon horn, not a bugle), breakfast at 0545,
and work started at 0615.

At Mount Tom Price, where Hamersley Iron Pty. Ltd. are
excavating iron ore, Taylor Woodrow International have
installed a crushing plant, together with tracks and conveyor
installations. Nobody knew there was iron in this almost
unexplored range until 1952, when an Australian owner of
asbestos mines named Lang Hancock, flying solo from

7. City Arcade, Perth, a Taylor Woodrow development.

8. Power in the Andes. The Miraflores Dam, Colombia.

Wittenboom to Perth, dived low to avoid a storm and noticed that his compass needle had gone crazy.

Here dust storms are the main problem. In this barren waste a new town has been built, with a shopping centre, health, education, and sports facilities. 'A mountain of iron ore is being worked for export,' Brian Trafford reported. 'Everything is huge, blasts of up to 100,000 tons of ore being quite usual, with 100-ton trucks hauling their loads to crushing plants, and mile-long trains shunting the crushed ore north to Port Hedland.'

Taylor Woodrow International Ltd. now operates a mining services division, based at Applecross, South Perth. It offers a complete range of mining services, including aerial survey, geological investigation, exploratory drilling, shaft sinking, and open pit mining. Among recent contracts is one for the drilling of some 5,600 feet of eight and six inch diameter bores, carried out for the Mount Newman Mining Company.

Taylor Woodrow (Australia) Pty. Ltd. has also bought a 8,600-acre farm for wheat and sheep. Called Mundjuwiring (meaning, in Aboriginal language, 'a wooded row of trees' – 'Woodrow'!), it is at Dalwallinu, 150 miles north-east of Perth. This is a new kind of diversification for Taylor Woodrow, and Frank Taylor announced, towards the end of 1969, that the group eventually wanted to buy about five farming properties between Geraldtown and Albany, respectively about 300 miles north and south of Perth.

Farming seasons in the Antipodes are the other way round, of course. At the end of August, Peter Archer, the farm manager, calls in a fifteen-man team of shearers who shear 3,100 sheep in three and a half days. Fleeces, each weighing nine pounds, are baled into 350-pound woolpacks which are sold in Perth to foreign buyers.

The grain harvest is from the end of November to early

N

January, and about 3,200 acres of wheat, barley, and oats are harvested.

Frank Taylor visited Australia again in February 1970, with Sir Patrick and Lady Dean. They saw all the sites we have mentioned, and more besides, including houses at Cook Point and office buildings.

There is another new housing development, built by Taylor Woodrow Corser, at Wanneroo, near Perth. A new water system to supply the township had been installed and was ready just as the first residents moved in. It is an artesian bore to 930 feet from which water is pumped into a million gallon reservoir. Other new houses at nearby Lake Joondalup will also benefit from the water supply.

Taylor Woodrow housing developments are linked to the Australian Government's immigration policy. To see how it works, meet the Dyos family, John, Rita, and their two children Lorraine (thirteen) and Simon (nine). By skilful teamwork between the sales manager of Taylor Woodrow Homes (Tommy Fairclough, son of 'Tommy Builder') in London and Taywood Hughes in Sydney, a home for the Dyos family in Australia was arranged before they ever left London.

In October 1969, they had answered a Taylor Woodrow Homes advertisement in a London daily paper. Having saved a bit of money (£2,300) from the small newsagent's shop they had been running in Surrey for six years, they had already decided on emigration, and were waiting for a sailing date. Taylor Woodrow Homes went to see them with plans of housing estates in the Sydney district.

On March 5 1970, the Dyos family arrived in Sydney, and on March 10 in Wollongong. Within forty-eight hours they had selected a Taywood Hughes house, moved in, bought furniture, enrolled their children at schools, and found a job for John Dyos.

Mrs Dyos, of 82 Lake Entrance Road, Barrack Heights, Warilla, near Wollongong, N.S.W., could hardly believe her good fortune. 'Taywood Hughes saw the building society about a mortgage, arranged job interviews for my husband and helped me book the children into school. We badly needed a fridge, so Taywood Hughes's Wollongong office lent us the money till ours came through the bank. They also set us up with things like plates and knives and forks till our own arrived.'

Over to the west coast again. What is it like, working on civil engineering projects in Australia? 'Well, there's a shortage of skilled labour, of course,' JohnWilson says. 'It's an agricultural, stock raising country. We have to take a lot of our own people, and recruit Australians sometimes from the East side – *New* Australians, many of them. A typical work-force would be, say, sixty per cent British, twenty-five per cent Italian, and fifteen per cent Yugoslavian.'

Was there any recruitment difficulty? 'Less than you'd think. The word gets around, you know – Taylor Woodrow are known to treat their people well in U.K., and our reputation has gone before us. Teams who have worked with us are often willing to rejoin us on another site.'

And the future? 'So far we've played Australia by ear. We've seized some opportunities and created others. The pragmatic approach, you could say. Here in the outback, rapid, first-class engineering work is badly needed. For the moment, our work is likely to be concentrated on mining projects in the north-west, with Perth as our main base.'

And after that? John Wilson muttered a two-syllable word, and I think it was 'Queensland'.

XV

Go East, Young Engineer

OCOLNA, in south-west Romania, was once a country estate where the Romanian royal family and its friends came to shoot. A few miles to the south, the sluggish Danube acts as a frontier with Bulgaria. The nearest Romanian village, Amarestii, is six miles away; the nearest sizeable town, Craiova, is two hours' drive to the north. This is not the rugged Carpathian Romania of romance, but a vast rolling plain where peasants struggle for a living with crops of maize, sunflowers, and grapes threatened by the spectre of drought. You see them, fur-capped, driving bullock carts along unmade roads in clouds of dust, through straggling villages, each with its onion-spired church; a tough humorous race that has known much invasion and persecution through the centuries.

But the Romanian Ministry of Agriculture and Silviculture has been at work here over the past twenty years. Great belts of acacia trees stretch to infinity – windbreaks that also bind the soil against erosion.

In October 1970, when I visited them, about seventy British people were living at Ocolna, including twenty-four wives and fifteen children, with a contingent of Romanians and a number of gipsies in the adjacent village, who are very good about providing donkeys for the children to ride. Among the acacias and chestnut-lined avenues of the estate, an English village has sprung up, with roads called 'Orchard

Road' and 'Woodland Way' for the married quarters, and apartment houses named after Brunel, Faraday, Watt, and other great engineers, for the bachelors. By the time these words are read, there will be a clubhouse, tennis courts, a swimming pool, and an open-air cinema; and every house will boast a garden. There is already a gritty, improvised golf course among the engineering works.

This self-contained community has a Families Association, a Bachelors Committee, and a Club Messing Committee. It is visited every so often by the Rev. Derek Cordell, dog-collarless, tweed-jacketed chaplain to the British Embassy in Bucharest, 150 miles away, who has the care of all expatriate British souls in the Balkans. The Romanians provide a nurse, a surgery, and young Dr Bonescu from Amarestii, who is available for two hours a day. The commonest complaints are sore throats and catarrh, which may have something to do with those clouds of dust along the furrowed tracks.

Entertainments are mostly home-made, dances, dinner-parties, car rallies (heaven help your suspension on these roads) a Scottish reel group, and bright ideas displayed on notice boards: 'Will anyone interested in buying and racing go-karts please sign below.' Television, rather fuzzy, comes from Sofia; and you consider you are keeping up with things if you manage to get a London newspaper ten days after publication. Post offices are one of the things Romania is not good at.

How anyone has time for any recreation is difficult to see, since almost the entire population of the camp gets up before dawn and returns after dusk seven days a week. Meanwhile wives shop and cook and have coffee mornings, and the children, whose age range is five to eleven, go to Miss McKenna's school. Twenty-five-year-old Miss McKenna, who comes from Brentwood, Essex, took this job as a change from teaching deaf children and because she wanted to go abroad. Her methods are modern, and the children will never know what an inch or a pint is: they are drawing and measuring in metric right away: 'This is a picture of Miss McKenna's right foot, which is twenty-six centimetres long.'

The reason they are all here is T.W.I.G. – the Taylor Woodrow Irrigation Group, a consortium of Taylor Woodrow International, G.E.C. Electrical Projects, Sigmund Pulsometer Pumps, and Vickers, headed by Taylor Woodrow International director Michael Thomas and with, as London manager, Lionel Edwards, back in the United Kingdom after many years overseas. This team is constructing one of the world's biggest irrigation systems over an area of 300 square miles, twice the size of the Isle of Wight. When it is finished a man in a field of maize will turn a hydrant to produce artificial rainfall and this single action will activate automatically a sequence of pumps and weirs to replace the water used, and at the end of the line will open slightly the gate at the Danube bank to let more water into the system. The water comes out of the river and is pumped up to a height of 118 metres. The Romanians are highly excited about the project, which is already attracting world attention: there is hyperbolic talk of 'grapes as big as plums', but the main requirement is more than one crop a year of all the fruits and grains that Romania needs to feed herself, and which she hopes to export through barter agreements in exchange for capital goods.

Said *The Economist* on July 5 1969: 'This £23 million deal gives a hint of just how great is the potential for expanding east–west European trade when East European bureaucracy and West European import controls do not stand in the way. This deal is just the sort of self-financing kind that East Europeans love. Now that the Danube is being increasingly brought under control, much of the massive flood plain and the marshes surrounding the lower reaches have been reclaimed for agriculture. . . . Agricultural expertise will come from the British Agricultural Export Council.'

Expertise also came from Dr Binitsa's experimental farm near by, and from the Agricultural Research Institute at Bechet, in whose outhouses the Taylor Woodrow team roughed it during their first winter there. Deputy Minister of Agriculture, Barbu Popescu, looks forward to a time, only a few years away, when his fifteen-year policy of 'Mechanize-Irrigate-Fertilize' will have transformed south-west Romania. He reckons that from 1971, the irrigated area will increase by 250,000–300,000 hectares a year, so that by the end of the next five-year plan the irrigated land will cover nearly 2·5 million hectares (more than 8,600 square miles).

The Taylor Woodrow irrigation area is partly on the terraces of the Danube and partly on a reclaimed flood plain. One of the first tasks to prepare the area for the project was the drainage of Lake Potelu (about twelve miles long, and a much missed source of fish for the local population, who must console themselves with the prospects of no more floods and two peach crops a year instead).

The water is distributed by means of buried P.V.C. pipes supplied by 134 automatically controlled pressure pumping stations, each serving a plot of about two square miles. The water is brought to the pumping stations by a system of lined gravity canals with automatically operated control weirs. Each pumping station is equipped with distribution pipework ('antennae') and sprinklers to give exactly the right amount of 'rainfall' for a range of crops.

From the Danube, the water comes through an automatically controlled gate ('head regulator gate') through a low level canal across the reclaimed flood plain to a main pumping station, designed to lift 43·5 cubic metres of water per second to a height of 71 metres through a 5·17 km steel conduit (known as a 'penstock') of 3·65 metres diameter, which discharges into an open-lined canal just over 7 km long.

From this canal a second main pumping station lifts 24·5 cubic metres of water per second to a further height of 24 metres, through another steel conduit 2·8 metres in diameter and nearly 3 km long. This in turn discharges into another open-lined canal just over 6 km long. Canals and conduits feed a system of distribution canals (about 190 miles of them) lined with polythene sheet protected by precast pressed sand-cement slabs, hexagonal and interlocking.

The working of the whole system is constantly monitored at a project data centre.

The sixty-seven square mile flood plain can be irrigated either by canals or by boreholes; but the flood plain canals are unlined because they have the double function of water supply and drainage.

Back in 1969, just after the Anglo-Romanian deal was clinched, big grey transport lorries with ROMANIA on them became a familiar sight around Taylor Woodrow's Ealing and Greenford premises. They belonged to Romtrans, the Romanian Government Transport Agency. They had unloaded fruit in London, and they were now loading factory-made components of Swiftplan houses to build the Ocolna township for the T.W.I.G. team. By continuous driving across Germany, Austria, and Hungary they reached their destination in four days, for the township had to be built before the onset of winter.

Meanwhile the first £150,000-worth of earth moving equipment, assembled from Scotland, Wales, East Anglia, and the West Country, was converging on Manchester Docks, where the Romanian MV Galati was waiting. The shipments reached the Black Sea three weeks later, and took a further three weeks to reach Port Bechet, on the edge of the site area, by Danube barges. They included two twenty-three-ton earth scrapers, tractor shovels, articulated bulldozers, hydraulic excavators, motor graders, and vibrating rollers.

Something odd was happening at Greenham's Hithermoor Pit, Staines, when Mr Popescu and a Romanian Government party came to see it in March 1970. A 100-metre length of prototype irrigation canal had been constructed, with sluice gate and pumping station – an exact replica of what was going to happen in Romania. It was the interlocking slab system that fascinated the visitors, for it was believed to be the first time such slabs had been produced by hydraulic pressing, and also the first use of sandy soil as aggregate for this purpose. A prototype pumping station with weir gate was also built for testing studies.

On the site, two special factories have been built: one for making penstock sections, the other for making concrete slabs. The penstock sections are made by bending cold steel half an inch thick into cylinders; the slabs, made of concrete mixed in a Hotcrete machine, are stamped out like biscuits.

Relations with the Romanian labour force (I.C.I.C., known as 'Itchik') are friendly, and until the language difficulty was overcome (Taylor Woodrow pay a £50 bonus to any member of the team who learns Romanian in three months) were based largely on football matches, which the Romanians, who practise three or four times a week, invariably win.

Problems arise through differences in technical sophistication: Romanian carpenters, for example, still use an all-purpose tool called in English an 'adze', and it isn't much good for making accurate shuttering for concrete. But tools imported from Britain would have to be paid for in sterling.

'*Problema, domnul!*' (Problem, sir!), always accompanied by a smile, used to be the most frequently heard phrase in the early days of the contract. Romanian, a largely Latin language with an admixture of Slavonic and bits of Turkish, comes fairly easily if you know a little Italian or remember some school Latin. I managed to direct a car driver from Bucharest to Ocolna by means of odd words recalled from Caesar's Gallic War.

Gordon Craig, the project manager, brought the nucleus of his team from his previous project, an airstrip in East Pakistan. They include Dave Lawton, a bearded Welsh construction superintendent who, like so many Taylor Woodrow people abroad, is a veteran of West Africa. When I was there Arthur Angell, a foundation expert, had just arrived from the Hartlepool nuclear power station as acting chief engineer. Mike Foster-Turner, works co-ordinator and planning engineer, and a genius at organizing entertainments, has spent much of his working life abroad. Doug Stevenson is a young Scottish pipelaying specialist. His wife runs the 'village shop'. Wives are immensely important on overseas sites – they help out as secretaries, direct operations in the kitchen during temporary crises when there is no caterer, and teach the children when there is no school.

Any project manager's job is a matter of human relations as much as of engineering and administration. It is he who has to reason with the young bachelor whose 'first time out' temporarily depresses him; who has to learn that a civil engineer's job is like this, and that, backed up by a friendly community, it can be a robust and enjoyable life. It is the project manager who may have to decide that X is temperamentally unsuited to this particular site, and recommend his replacement. It is he who gets up before anybody else in the morning, and is still visiting the sick last thing at night before he goes to bed. It is he who has to lead negotiations with, in this case, the Romanian construction organization; he who must sense at once if there is any discontent in the camp, and call a meeting or appoint a committee to sort things out.

In Romania it is he who, suddenly told that the Minister

of Agriculture is due to arrive in two hours' time, alerts everybody in that 300 square mile area through T.W.I.G.'s twenty-four radio stations (most of them mounted on Land-Rovers) – Harold Minto at the slab factory, Giovanni Pensa and his dredgers, Alastair Sutherland in charge of shuttering at the main pumping station (called L.1 – L for 'lift'), the Vickers boys at the penstock factory, Mac Morum on the Eastern Terraces. . . .

Out there, in the irrigation area, small thatched homes survive between the canal excavations and shepherds and goatherds guard their flocks against these new hazards. Terex scrapers, groaning like huge wounded rhinoceroses, scoop out the forty metre wide main canal: when they have gone deep enough, the dredgers will take over.

It won't be long now: precious water and crops of maize, winter wheat, tomatoes, tobacco, sunflowers for oilseeds, will turn these sandy plains into a land of Cockaigne. And this is only the first stage. A northern extension was being discussed while I was there – another 55,000 hectares or 210 square miles, on which work might begin in 1972, using the existing penstock and slab factories, and many of the same team from Ocolna.

We have so far shown the civil engineer as a pioneering character hacking order out of chaos, often in acute discomfort. Just occasionally, however, he finds himself in a station where he is the envy of all other construction teams the world over. John Bevan's team, who are building container wharves in Singapore, live in houses and flats near the Botanical Gardens and drive to work like gentlemen. Once there, they put on hard hats and strip down to their khaki shorts. They have a leave bungalow by the sea near Kuala Lumpur, about seven miles from Port Dickson; and there is an authentic British pub, in Tanglin Road, to relax in.

Lee Kwan Yew's Singapore, an island the size of the Isle

of Man, ninety miles from the Equator, has a country club with four golf courses, a slum Chinatown which will soon be cleared, and great industrial plans for the future. It is a multi-racial society that seems to work. Nobody has a chip about being Chinese, Malaysian, Japanese, American, Malay, Pakistani or British. Half the two million population are under nineteen, and per head prosperity is second only to Japan in Asia. The city is so clean that you will be fined thirty dollars if you throw your cigarette end on the pavement in this, the world's fourth greatest port.

Off-shore, near the faint outline of the coast of Sumatra, you see oil rigs – just a hint of the huge wealth of petroleum, on-shore and off-shore, that is known to exist in South East Asia among the Indonesian Islands, and one of the reasons for Singapore's future importance; and a few yards from Taylor Woodrow's site offices is a relic of Britain's imperial past – the original Royal Singapore Yacht Club, now occupied by the chief engineer of the East Lagoon development.

Four new berths are needed to bring Singapore into the container age. They are being built in the East Lagoon, from land much of which has been reclaimed. The first 'feeder' berth (Cross Berth 48), 700 feet long and 300 feet wide, will provide berthing in thirty-two feet of water at low tide. It was completed exactly three months ahead of schedule, fourteen and a half months after the first pile was driven on June 16 1969, and officially opened by Mr Yong Nyuk Lin, Minister of Communications, on November 4 1970. It was he who had pressed the button to drive the first pile.

On this occasion, the Taylor Woodrow International team was represented by Bob Aldred, chairman, and John Bevan, project manager. Congratulating the construction team on their 'very creditable performance', the Minister remarked that they must all have been keen stamp collectors, since they had completed the berth in time for it to be shown on three commemorative shipping stamps issued on that day.

Once fully containerized ships are introduced on the Far East route, Mr Yong Nyuk Lin predicts, as much as twenty-five per cent of general cargo handled in Singapore will be containerized. These container wharves were planned as long ago as 1959, since when containerization has developed faster than anyone foresaw. Mr Howe Yoong Chong, chairman of Singapore Port Authority, estimates that about four million tons of container ships are now under construction.

Container ships grow bigger and bigger: the latest are 950 feet long, and Taylor Woodrow may be asked to make the wharf longer than the projected 2,250 feet, which gives enough room for three ships – but it may take only two.

Cross Berth 48 can accommodate 1,000 containers, Berth 49 will accommodate 3,060 containers, and Berth 50, 2,134. There is provision for eighty-two refrigerated containers, and twenty-four more if required.

The completed wharf deck will be supported on over 7,000 concrete piles, which, placed end to end, would stretch for ninety-five miles.

The design is for heavy concrete piles made on the site. Each hollow-spun cylindrical pile takes about seventy-five minutes to make. Steel bars are coiled for 'helical reinforcement', threaded with longitudinal bars, and filled by hand with concrete. The whole is then placed in a mould and spun at up to 300 r.p.m. for thirty minutes. These 110-foot helical piles are probably the longest ever made without joining. Beams are cast on a 'long line' system, using high-tensile strands twisted like a cable.

The Taylor Woodrow expatriates, seventeen in all, work with a peak number of about 540 nationals and sub-contractors. Everyone is on site from 7 a.m. to 6 p.m., six days a week: because they are on the Equator, there is no winter problem of the 'days drawing in'. 'We have the enormous advantage of Singapore workers,' says John Bevan, the project manager. 'They actually seem to *love* work, they appreciate our rates of pay and they are very quick to learn new skills.'

John Bevan joined TaylorWoodrow Construction in 1955, spent six months as site engineer on Nottingham power station, and a year at Northfleet. This is not his first harbour – he was chief engineer at Port Harcourt for three years, and construction superintendent at Pepel jetty in Sierra Leone. London Airport claimed him as chief engineer for its second multi-storey car-park, and he was then project manager at Jounieh Port, fifteen miles north of Beirut. In between these overseas assignments, he was a project engineer at Western House on tendering, with trips abroad for site investigation.

At Pepel, fourteen miles up a muddy creek from Freetown – 'a typical bush job' – John Bevan worked with other veterans of West Africa, including John Calkin, who is now chief engineer on the Singapore project. Roy Tibble, quantity surveyor at Singapore, was also at Pepel and at the rutile mine, of which his chief memory is 'snakes'. He was, too, one of the long-suffering Taylor Woodrow team at Marsa el Brega, Libya, where they were interrupted by the Israeli–Arab war: the Arab workers left in a body to join up, just when the huge concrete caissons, which had been made in Sicily, were being towed across the Mediterranean to the site. Walter Hogbin, in charge of landing operations, routed out all hands (including the office staff) to man the winches and haul the caissons into position for sinking. Walter, a Cambridge engineer in his thirties, is now at Singapore too, and has also been at the Sizewell and Wylfa nuclear power stations, where incidentally, he met his wife – many construction men meet their wives this way.

So you thought the Mediterranean was a tideless, peaceful ocean? There were twenty-two-foot waves, capable of knocking a fifty-ton crane over – and £60,000 worth of equipment was lost in a storm while it was being towed across on barges.

This was what Americans call 'grassroots engineering' – 100 expatriate staff and no wives: they even had to bore their own wells and distill all water. But there is hardly any member of a Taylor Woodrow overseas team who cannot tell stories like this. Gil Adams, section superintendent of wharf superstructure at Singapore, was in Ghana in the middle of the anti-Nkrumah *coup d'état*; and Frank Westaway, plant supervisor at Singapore, recalls the furious mockery of baboons as he and his men tried to clear hardwood rain-forest at Guma – first with axes, then with fire, and finally explosives.

Last man out of Marsa el Bregha – he was actually under house arrest for several days – was Duncan MacIntyre, now in Singapore in charge of administration of the nickel project, 900 miles away in Sulawesi (Celebes), for I.N.C.O. (International Nickel of Toronto).

In the first chapter, we gave some hint of the difficulties of setting up a camp in equatorial country with almost no communications. The first man ashore at Sulawesi, nearly three years ago, was fifty-seven-year-old Jack Cooper, ex-Coldstream Guards R.S.M. He and a colleague arrived by *prahu* up-river from Malili to build the first camp, a hut (with mosquito nets) to house eighteen men. From the nearest village they hired a girl to make bread, another to clean. Stores arrived; then a laboratory, then the materials for a helicopter hangar – these were the priorities. The local people who had seen no Europeans since the old Dutch colonials, entered into the spirit of the thing and gave every help. Not for four months did the first British wife come out.

There is a mixed European technical staff; but all the

geologists are Indonesian. From the main headquarters, field camps are established by sending a handful of men off into the blue with a bulldozer and a drilling machine. They recruit labour to build round rattan huts to live in; and eventually they send out a geologist alone with his drill to establish a 'fly camp'. In these conditions, much may depend on a man's ability to improvise, and the name of Les Dimond (younger brother of Roy Dimond, plant manager on the Singapore contract) is often mentioned; it is he who can make a spare wheel for a vehicle, or design a sledge for muddy ground; he who has a gadget for every emergency.

Once soil samples have been taken (they are known as 'gunk'), they are handed over to I.N.C.O., who alone know the results. This secrecy is deemed to be necessary because several different countries, including Japan, are probing here, and the Indonesian Government does not want a 'nickel rush' with claim-staking and inflated land values.

Headquarters camp is now in full working order, and more British wives have arrived with children. They live in Arcon houses on concrete bases. Since there is as yet nobody to teach the children, Duncan MacIntyre is arranging for the older ones to go to boarding school in Singapore. Seventy kilometres of earth road have been built to link the base camp at Soroako with the Malili headquarters. Taylor Woodrow have their own landing craft, the good ship *Ever Constant* with ten Indonesian crew. They transfer incoming cargo to barges, which bring it nine miles up-river to Malili.

Life at Malili, after many hardships, is now almost luxurious, with concrete tennis courts, radio, tape-recorders, a library, bicycles for the children – all supplied from Singapore by Duncan MacIntyre, his Chinese girl secretary, and Mr Lim, his driver/clerk.

At Christmas, Santa Claus arrived at Malili by dumper-cart, bringing toys which had been ordered from Singapore's lavishly stocked shops the previous September. And life can feel pretty good when you see it from the 'Soroako Hilton' – the Taylor Woodrow Club built on stilts over a beautiful mountain lake 2,000 feet above sea level.

Back in Singapore, there is one more enterprise that should be noted – and it is intimately connected with the future development of Indonesia. Ralph King is managing director of Arcon (Singapore) Pte. Ltd., a subsidiary of Taylor Woodrow (Building Exports). In the past three years, under managing director Arnold 'Tommy' Thomas – who had joined the company in the post-war Arcon housing drive – Building Exports has not only identified itself with Prime Minister Lee Kwan Yew's dynamic industrial policy in Singapore, but seems to have discovered an almost unlimited new market for factory-built structures in South-East Asia and the Pacific islands.

In the south-west of Singapore, the vast Jurong Industrial Estate (the largest of several such estates currently being financed by the Singapore Government) is being created out of 16,000 acres of hill and swamp country – they are turning it into dry land by pushing the tops of the hills into the swamps. This is hailed by Singaporeans as 'new frontier' territory – a man-made land of factories, workers' houses and flats, railways, deep-water berths, and road, telephone, and power networks. It is the Singapore of the future.

Arcon was in on this very early, and provided the first fourteen pilot factories. So another development has resulted from the original Arcon structure – Neel's wartime idea in Britain for using the few materials that were available (such as tubular steel) in times of shortage.

Arcon also built the hall at Kuala Lumpur in which Malaysia's independence celebrations were held in 1957, and was able to put up 100 buildings for the British Army in Brunei in record time. The structures are so simple to erect

o

that the boys of the Montford Boys' Home, Kuala Lumpur, built the Home themselves!

But it is the prospects in Indonesia that excite Ralph King. Three very large government textile mills, in Bandung and elsewhere, have been built under 'package deals', with Taylor Woodrow constructing, providing equipment and supervising its installation. Milk processing and pharmaceutical factories, assembly plants in Java and Sumatra, schools and other buildings designed to order, steelwork and accessories, doors and ceilings – it looks as if there may soon have to be more Arcon companies in this vast area.

'I would say that about sixty per cent of our business here is industrial, and forty per cent amenities of various kinds,' said Ralph King. 'At the moment' (he was speaking in October 1970) 'we have over a million square feet of new business in South-East Asia, most of it in Indonesia – and somehow it has all got to be completed in four months!'

XVI

The Team and the Public

A CONSTRUCTION project can have a pretty violent
effect on a community. A motorway is driven through
your village. Traffic is diverted in a city because of a
yawning hole in the ground. Ugly machines and tough,
helmeted men appear. Construction men are both seen and
heard, and they cannot help it. One day the whole site will
settle down into a skilfully landscaped hydro-electric dam, or
a reasonably graceful tower-block. But until it does, we have
to live with the construction men. We do wrong to blame
them for 'the noise and the people'; for behind them are the
planners whom we seldom see.

The construction industry in recent years has developed a
keen sense of social responsibility. It uses the word 'environment' much oftener than it used to, and in the last chapter we
shall try to predict the future of this attitude. It has a slight
inferiority complex about being seen, by the uninformed
general public, as a bunch of 'navvies with string round their
trousers'. It nowadays goes out of its way to maintain good
relations with the public, and it is fair to claim that Taylor
Woodrow have led the way, and that this policy emanates
from Frank Taylor himself.

We have seen, in an earlier chapter, the importance the
group attaches to good internal communications and
human relations. Where Taylor Woodrow are in advance of
many other companies is in their recognition that internal

and external public relations are two halves of the same thing.

It is done with both humanity and humour. A nocturnal demonstration of the Pilemaster at Euston Station was accompanied by the playing of *Silent Night*. The famous public observation platforms are at once a benefit to the industry and to the public, and also a bit of a joke. As Frank Taylor said in one of the numerous press and newsreel interviews that followed it, 'everyone is a builder at heart. The public go to extraordinary lengths to see what is going on and we feel we would like to make things easier for them.'

The story even reached the 'Live Letters' column of the *Daily Mirror*, in which a Mr J. Winter, of Woburn Square, W.C.1, wrote: 'Three cheers for Mr Frank Taylor and his pavement supervisor's observation platform. . . . I love to applaud while watching a neat, nifty manœuvre on a bull-dozer, or a terrific solo on the pneumatic drill. Now that Mr Taylor has set the example, the gallant gentlemen of the building trade can take the curtain calls they so richly deserve.'

Taylor Woodrow also pioneered the policy of 'apologetic notices' at difficult sites, expressing genuine regret for the inevitable inconvenience they were causing the public. In social life this would be elementary good manners, but not every industrial organization would think of it.

The group are conscious of their responsibilities towards the trades (and their unions and managerial associations) which go to make the construction and civil engineering industry, and Taylor Woodrow men at all levels are active in them. Frank Taylor in particular has always had a deep feeling for the traditions of the building trade, and the honouring of its traditional customs and rites. These began, in ancient times, with human sacrifice: Holsworthy Church, in Devon, is said to have been built over a living human being.

This practice being both obsolete and illegal, others are observed all over the world – turning the first shovelful of earth, putting coins and newspapers in cornerstones, celebrating the roof-level of a building. In Italy, it is still the custom to tie a sprig of fir tree to the top of the scaffolding.

'Topping out' is the most universally observed rite today. If beer is drunk, it tends to be known as 'wetting': the brewers invariably do this when a new pub reaches its full height. But it must be done *at the top*, even if you suffer from vertigo.

If you stop anyone in the street and ask him: 'What does Taylor Woodrow mean to you?' he will almost certainly say 'four men'. All contractors display large signs on boardings and cranes, usually floodlit at night. But none is so universally familiar as the symbol of four men pulling on a rope, signifying team work.

The design was developed from the winning entry by Maurice Higham, a storekeeper of Taylor Woodrow Plant Co. Ltd., in a group-wide competition, and chosen from hundreds of entries. In 1958 a monolithic granite sculpture, based on the 'Four Men' design, was commissioned from David Wynne, who has specialized in sculptures of sportsmen and athletes. Eighteen feet long, seven feet high, and four feet wide, it now stands in front of the façade of the Taylor Woodrow office block at Hanger Lane, Ealing, headquarters of the overseas companies, and of Taylor Woodrow Homes, Myton, and Phillips Consultants. It weighs over twenty tons and is believed to be the largest monolithic granite sculpture since the days of ancient Egypt.

A natural by-product of the symbol was the Tug of War Association's invitation to Frank Taylor, in 1969, to become Patron in succession to Sir Billy Butlin. Since 1966, Taylor Woodrow have sponsored the annual National Indoor Tug Championships at the National Recreation Centre, Crystal

Palace, at which teams from all over England compete, and a team from the Taywood Sports and Social Club generally takes part. Taylor Woodrow were again sponsors of the European Tug of War Championships in September 1970, and they have also donated a trophy to the Australian Tug of War Association in Sydney.

Wherever it is working, the group tries unobtrusively to identify itself with the life of the district or country it is in. This is not always easy: many a project manager has been known to say that his job, on an overseas site, is as much public relations as civil engineering. Nearer home, it is easier: when Taylor Woodrow were building the Lednock Dam in Scotland in 1956, they revived the Comrie Highland Gathering after twenty-four years, and presented a cup for caber-tossing which was won by (who else?) the great Ewan Cameron, landlord of the Lochearnhead Hotel. And in August 1969, on the opening day of the Nottingham Summer Race Meeting, the Victoria Centre Development was commemorated by the New Victoria Centre Cup race, sponsored by Taylor Woodrow Construction (Midlands) Ltd. and their partners.

On site, a construction company often finds itself in a position to be a good Samaritan. It has cranes to lift crashed cars, and because of its own vigilance about safety, it can provide medical equipment. Klaas van der Lee, project manager on the Invergordon contract, received this letter from the Chief Constable of Ross and Sutherland:

'I am writing to thank you for the kind assistance rendered by your Mr Charles Parker, a member of your First Aid Post team, at a serious road accident at Dalmore Bridge, Alness, on August 27 1969. I understand, too, that this gentleman arranged for the supply of blood plasma, and there is little doubt that its timely administration helped greatly to relieve those who were seriously injured.'

At Sizewell, in August 1961, prompt action by four team members saved the life of an American Air Force pilot, Captain Philip C. Gast, who had ejected from his crippled aircraft over the sea.

Roy Broadhead, Dave Woodard, and Harold Powell, civil engineers, and Eric Rothery, foreman fitter, saw the plane plunge into the sea about a mile off-shore. They immediately jumped into a fibreglass boat and made their way through rough seas to the pilot who, due to a back injury, could not collapse his parachute, which ballooned and pulled his dinghy farther out to sea. The rescuers were forced to steer their boat on a collision course to intercept. Somehow they fought their way back to shore, and the pilot was treated at Taylor Woodrow's Sizewell medical centre.

The Squadron Commander, in a grateful letter to Vic Blondell, then superintending the nuclear power station construction, said that the rescuers' accomplishment 'reflected great credit upon themselves, your organization and your country'.

David Coppin's George Medal was won for a very different piece of heroism. David, then a personnel manager at Western House, saw four armed raiders enter the Westminster Bank at Hanger Lane in February 1966. He pushed their getaway car against a concrete pillar, tackled all four and captured one of them. They were all eventually arrested and received sentences up to five years.

Mutual concern for safety between the industry and the public is the reason why Taylor Woodrow display 'Accident Record' figures at some sites – even at Liverpool Metropolitan Cathedral in 1962.

'Safety' is often the theme on which agents or project managers base their own local public relations. When Taylor Woodrow were building the Aston Expressway, the seven-lane 'tidal flow' artery connecting the Midland Links with Birmingham city centre, the agent, Jack Monger, found a hostility among local schoolchildren which was expressed in vandalism and pilferage. So, with his director, Frank Gibb, he devised a programme of safety competitions, for essays and paintings, in local schools, which were also given safety lectures and – for the younger children – 'adventure playgrounds' which a construction company can provide very quickly out of odds and ends of material.

Construction companies are not 'big spenders' in the advertising league, compared to manufacturers of 'consumer durables' such as cars.

Most leading civil engineering companies place 'corporate' or 'prestige' advertisements in newspapers and periodicals known to be read by the managements of potential clients – the people to whom they are really selling. These usually indicate the large range of services offered; or they may particularize, to show their experience in building, say, schools. Lately, as construction groups decentralize into the provinces, there has been a tendency to say: 'We're not as big-headed as people think – we're interested in five-figure contracts as well as eight-figure ones.'

Taylor Woodrow are very active in press relations, which are seen as an essential part of the whole publicity picture, and also make documentary films of their outstanding achievements. Paragraph 33 of the confidential Company Policy document, circulated to all executives, says: We must endeavour to keep our clients, shareholders and employees, and also the general public, well informed of our activities. This is done by means of correspondence, advertising and editorial publicity in the national and technical press.'

Nat Fletcher, group director of publicity, considers advertising, marketing, and public relations as a whole when he says:

'Publicity has other objectives as well as sales promotion,

as we also want it to have a special message to investors, bankers, suppliers, and sub-contractors, to assure them that our organization is worthy of their confidence. We spend a lot of time and thought to try and ensure that our publicity will promote a feeling of pride in our organization among members of the team. . . . Nevertheless, it should be emphasized that every team member is an ambassador for the company. The finest and most effective advertising of all is a job well done, but the more this is known and talked about, the better for us all.'

For shareholders, Taylor Woodrow produce one of the liveliest, most readable and best illustrated annual reports, with Frank Taylor's own personal touches in abundance, and invariably a symbolic picture of himself in his safety helmet. You really feel, as a shareholder, that you are participating, that you, too, have poured your bit of concrete as a team member.

The industry as a whole is somewhat bedevilled by the 'speculative builder' stigma. Yet it can be argued that the Adam brothers at Adelphi were speculative builders; so were Richard Grainger and John Dobson of Newcastle-upon-Tyne, Ralph and John Wood at Bath. And the conventional image of the contractor as 'pioneer and gambler' is inevitable, since in order to retain a leading place in any industry, a firm must break new ground and take on tasks needing skills and techniques not yet proven. For 'gambler' read 'calculated risk taker', and it makes better sense.

It is generally conceded that Taylor Woodrow have a clearer public image than some of their rivals. The small beginnings – two houses in Blackpool – are remarkably widely known; so is the personality of Frank Taylor; so is the fact that the group build power stations. Yet other big firms have also built nuclear power stations, McAlpine among them. Most of Taylor Woodrow's rivals have been in business longer (McAlpine celebrated their centenary in 1963), and most have dramatic pioneering stories to tell. If there are particular areas in which Taylor Woodrow can claim a special reputation, they are undoubtedly research, advanced design, and spectacular large-scale engineering, often in far-flung places.

At the consumer end of things, Taylor Woodrow Homes offer a personalized service which gives the buyer a real sense of welcome. 'We don't just build houses – we provide homes' is no mere slogan. Managing director Don Slough says: 'Some of our purchasers are now moving into their third Taylor Woodrow home.' It is not only that the design has been thought out to the last detail, but so has the move itself – a dustbin and clothes-posts await you, and, to your surprise, so do a free delivery of groceries, a vase of flowers, a bottle of champagne, and a packet of change of address cards.

Frank Taylor was always somewhat ahead of the times in publicity. Even in the 1930s he would invite Elsie and Doris Waters, then at the height of their popularity as radio stars, to open a show-house on a new estate in the Midlands. The idea was new then. In recent years, publicity has tended to become public relations – a subtler thing that has much in common with good manners.

XVII

Building for the Future

CIVIL engineering has been defined as 'the art of directing the great sources of power in nature for the use and convenience of man'. In this last chapter we shall try to look at its future, and at the place of Taylor Woodrow in that future.

A century ago, an engineer like Brunel could point to a whole project and say with pride, 'I did that.' Now we have specialization, and the risk of fragmentation. You cannot build a power station, a dam, a steelworks, a factory, without calling upon an enormous number of skills. Hence the emphasis on, indeed the absolute necessity of, teamwork.

We live in a Marshall McLuhan world of frightening speed. 'What the civil engineer is doing,' wrote H. J. Booth in *The Guardian* of July 9 1969, 'is to use the fruits of intensive specialization to create works that are bigger, safer, and cheaper, and complete them at a much faster rate.' Man, he said, is not only moving out into space, he is also moving back into the sea with off-shore mineral exploitation, pollution control, land reclamation and dredging, and under-water engineering; and, as an example of vast transformation of the earth's surface, he quoted Taylor Woodrow Irrigation Group's irrigation project in south-west Romania.

'The world-wide team of engineers, constructors and developers': this is how Taylor Woodrow today describe themselves, and the word 'developers' is a portent for the future. Allied to it are the words 'urban renewal' and 'environment'.

It was Norman D'Arcy, a non-executive director of the parent company, who persuaded the board to expand into property in a big way. So, in 1964, Taylor Woodrow Property Co. was formed, with George Dyter as managing director. There are critics who have said that the group should have done this much earlier, at the beginning of the boom; but it is a world that calls for caution. A paper by Colin Hunt, now managing director of the property subsidiary, to the main board in 1963 contributed to the new thinking.

Colin Hunt, a chartered surveyor and architect who was formerly with Hammersons, and his deputy, John Topping, are the authors of the £25,000 *Fulham Study* on urban renewal, the first and only study of its kind since the war. This was the result of an invitation to Frank Taylor by the Minister of Housing and Local Government in January 1963 to carry out a study on the possibilities of large-scale redevelopment of obsolescent residential areas in London, on the basis of partnership between private enterprise and local authorities.

It tries to avoid the muddle due to overcrowding in small out-of-date houses, to universal car ownership, to piecemeal short-term improvements by individual owners. By overall redevelopments, population density could be three times what it is now – with space for recreation, parking, schools, separation of people and vehicles by means of upper-level pedestrian decks. Public opinion is not yet prepared for such wholesale planning, and the study has not yet been implemented. It remains, however, a pointer towards the world we are about to enter.

'Property is complementary to building,' Colin Hunt says, 'because it brings in more building business.' American builders have been in property up to their necks for a long time: it is often good to *own* the buildings you build. Moreover, property ownership can underpin a construction company if new business should fluctuate.

Britain, a crowded country which has suffered the mass demolition of war, has long experience of urban renewal, and this has given Taylor Woodrow a know-how that has been found valuable in America and elsewhere – currently in Brussels and Perth, Western Australia, as well as at home. The day of speculative development may soon be over: social responsibility, and the environmental implications of what we build, seem about to take over. It can even involve creating employment in an area of unemployment or exporting whole industrial estates to the other side of the world.

The future is already being born at St Katharine Docks, just east of the Tower of London. Once a triumph of Victorian commerce and design, they are now a collection of derelict warehouses and waterways, used occasionally by film directors as sets for Battle of Britain films. Something very exciting is to happen here in the next decade – a £30 million redevelopment, to include a big hotel, an export centre, housing, schools, a sports centre, and a yacht club. 'It could give ideas to another city with moribund waterfront areas – New York,' said the *New York Times* in March 1970.

The area covers twenty-five acres, ten of which are water. Eastward along the Thames are nearly 800 more acres which may soon come up for redevelopment. Two of the existing warehouses, good examples of nineteenth-century commercial architecture, will be preserved and converted to new uses – housing, exhibition space, and (very British, this!) a yacht club (for sailing is no longer a rich man's hobby). The idea of a marina is believed to have been in the mind of Desmond Plummer, leader of the Greater London Council, since 1960.

The Greater London Council held a competition to pick a general design and suitable developers, allowing only three months for submission. It was won by Taylor Woodrow, who had Andrew Renton as chief architect and Ove Arup as engineering consultant, and Michael Young and Peter Wilmot, well known sociologists, among the advisers.

The export centre would be connected to the new £6 million, 830-room hotel, with restaurants and a floating night club. There would be both private housing and municipal flats; a sports centre with gymnasium, swimming pool, and sauna baths; a primary school and two nursery schools; and an inter-denominational chapel on the site of an old Hospice of St Katharine. The whole highly imaginative concept, blending renewal with preservation, was praised by the *Architectural Review* because it was obviously not made purely on the grounds of financial return.

The co-ordinating imagination was that of Peter Drew, a forty-three-year-old urban renewal expert and a director of Taylor Woodrow Property Co., who had to hold together a committee of theatre people, local government representatives, hoteliers, sociologists, and others, all highly enthusiastic but pulling in different directions.

'What mankind wants as much as peace,' he says, looking forward twenty-five years and more, 'is city renewal. Soon half of the population will live in cities. We've got to make sure we don't build future slums. We haven't got much room, so we have to recreate social urban life on decaying sites.' In saying this, he makes a careful distinction between 'urban renewal' and 'redevelopment'.

It is reported that when Frank Taylor saw the research brochure containing the whole plan, he said 'yes' after half an hour. There were hardened property men who were not so sanguine: 'No hotel group will ever build on that site! It's too far from the West End.' The answer came from Lyons' Strand Hotels, who had obviously studied the Greater London Council's and Corporation of London's future traffic-flow plans: 'But it's so *handy* for the West End!' Or as Peter Drew puts it: 'We're going to build a new West End in the East End.'

One hundred and twenty children are born into the world every minute. It is reckoned that world population growth over the next eight years will equal that of the first 1,800 years A.D. They will need food, houses, services, transport – especially in 'developing' countries. Land clearance and irrigation, on a scale and at a speed never before imagined, will be required. (T.W.I.G.'s 300-square mile scheme in Romania is just a taste of it.) So will versatile new machines, some capable of working on land or sea or swamp. Russell Jones, in the *Contractor's Plant Review* of November 1969 foresaw 'a new generation of crawler tractors', 150-ton dump trucks, eighty-ton 'tree crusher' tractors, capable of clearing fifty acres a day.

The demand for water will mean new dams and reservoirs. The demand for all forms of energy will need hydro-electric and nuclear power: in this industry Britain leads the field, and Taylor Woodrow's unrivalled experience in design and construction may be expected to play a part.

Stepping aside into science fiction, we may glance back at an exhibition at the Design Centre in London in November 1969. The idea was to show the impact of electronic communications on business in the year 1990. The example chosen was an office for a civil engineer, who, like most executives twenty years hence, works at home.

He has a 'vision-phone' to allow face-to-face discussions and inspection of details on the working site or documents. He can attend a meeting 8,000 miles away without leaving his study. He has push-button access to computers and telex machines.

The men who, stripped to the waist, are pioneering vast projects somewhere along the Equator will not perhaps take very kindly to a contracts manager who never visits the site. But perhaps by then it will be possible to operate radio-controlled bulldozers from the leafy suburbs of London and Manchester.

On the face of it, the construction industry cannot fail to take advantage of the expansion of demand we have tried to visualize. But it is still a fragmented industry (there are 80,000 contractors of all sizes in Britain) and it has been calculated that by A.D. 2000 sixty per cent of the world's business will be conducted by only 200 companies. What are the trends that point towards the civil engineering world of 2021, when Taylor Woodrow comes up to its centenary?

Frank Taylor, in his lecture to the Royal Society of Arts in 1963, was already forecasting 'the need to remedy the defects in so many of our urban centres now no longer well-adapted to the conditions of modern life'. Overseas, we had 'but scratched the surface of the opportunities to serve the new emergent nations'. There were also 'the challenge of new shapes and of new materials', and 'the rapid development of factory made buildings, with a whole room formed as a unit'. There would be 'a larger volume of work being carried out by smaller mechanized labour forces, but there is no reason why the numbers employed should be reduced – merely redistributed'.

Once we used to send pioneering young men overseas to build the Empire. Now we send civil engineers, in partnership with other countries: they wear helmets instead of topees, and they don't dress for dinner in the jungle. There is no doubt that international competition is growing keener, and as Bob Aldred points out, 'Twenty-five per cent of Taylor Woodrow International's business is in non-Commonwealth countries.' Nevertheless, 'by seeking work abroad, using local skill and labour where possible, the group spreads its risks'. Moreover, there is an increasing trend towards the 'joint venture' (among them those currently between Taylor Woodrow and the Dillingham Corporation in Singapore, Thailand, and Nigeria) to spread the risk still further.

The highly successful construction, on time and to scheduled cost, of the £37 million Invergordon smelter for British Aluminium endorses the trend towards the complete management or turnkey arrangement, which group companies offer on a world-wide basis. At Invergordon, for example, Taylor Woodrow and Head Wrightson's commitments covered not only their own extensive operations but the co-ordination and supervision of the contributions of some 2,000 manufacturers, suppliers, and sub-contractors. Plant worth £13 million was designed and ordered, and £2½ million worth of electrical equipment. Supplies came not only from Scotland and the rest of the United Kingdom, but from half a dozen North American and European countries.

Les Olorenshaw, joint deputy group chairman, who has been called Taylor Woodrow's 'administrative troubleshooter', sees fiercer competition abroad, especially in less developed countries; we may be up against Japanese, French, German, Italian, American, and sometimes Yugoslavian groups. Finance may be a critical consideration, as well as know-how: the Romanian irrigation contract was tied to a barter deal. The Common Market may bring more equity participation (as in the Grétima property development and estate agent business in Paris), using local capital and staff.

Whatever happens, it looks as if there will be no stopping the Taylor Woodrow group in the future. 'We could be five times as big as we are,' says A. J. Hill, the other deputy chairman. 'But there is no virtue in bigness alone. It is bigness of achievement we should be aiming at. But we must never, internally or externally, lose the personal touch. As we expand, I make it my business to have more and more

P

talk-ins with younger staff. We must always respect people. We must aim at total communication, person-to-person, across the globe.'

We make no apology for returning to the 'Taylor Woodrow Theme'. This document is circulated among all the group's executives, and in one paragraph it says all we have been trying to say in this book: 'Good systems do not automatically guarantee success. The most fundamental pre-requisite is that there shall be a team of competent people working together with intense and friendly spirit within the framework of sound and progressive policies. Such team spirit exists in Taylor Woodrow, and accounts for its present position and its prospects for the future.'

Illustrations — Acknowledgments

Colour

2. Wylfa nuclear power station, designed and constructed for the Central Electricity Generating Board by British Nuclear Design & Construction Ltd.

   Euston Rail Terminal, for British Rail London Midland Region. Taylor Woodrow Construction Ltd. was responsible for the engineering design and construction of the whole project under the direction of, and in collaboration with the Chief Civil Engineer of the London Midland Region. Architectural design was the responsibility of the Architect, London Midland Region.

3. City of London Headquarters Buildings, for the P & O Steam Navigation Company (left) and the Commercial Union Assurance Group. Architects: Gollins, Melvin, Ward & Partners; consulting engineers: Scott Wilson Kirkpatrick & Partners; quantity surveyors: Langdon & Every; mechanical & electrical design consultants: G. N. Haden & Sons Ltd.

4. H. V. I. Luboil Plant, Stanlow Refinery, for Shell U.K. Ltd.

   Cargo Tunnel, Heathrow Airport, London, for British Airports Authority – the project being under the general direction of its Department of Engineering. Design of tunnel, approach roads, bridges and associated works by Sir William Halcrow & Partners in association with Hoare, Lea & Partners for the electrical and mechanical services.

5. City Arcade, Perth, Western Australia – a Taylor Woodrow Development. Architects: G. Summerhayes & Partners;

consulting engineers: Halpern, Glick & Lewis; quantity surveyors: Rider Hunt & Partners.

6. Miraflores Dam, Colombia, South America, for Empresas Publicas de Medellin. Consulting engineers: Integral Ltda. of Medellin.

7. Midland Links Motorway, for the Ministry of Transport; consulting engineers: Sir Owen Williams & Partners; quantity surveyors: George Corderoy & Partners.

8. Liverpool Metropolitan Cathedral of Christ the King, for the Trustees of the Roman Catholic Archdiocese of Liverpool. Architects: Frederick Gibberd & Partners; consulting engineers: Lowe & Rodin; quantity surveyors: Franklin & Andrews.

Black and White

4. Atlantic House, Holborn, for Holborn Viaduct Land Co. Ltd.; architect: T. P. Bennett & Son.

5. Takoradi Harbour extensions, for the Ghana Railways and Harbour Administration; consulting engineers: Rendel, Palmer & Tritton.

   Bowater House, Knightsbridge, London, for The Land Securities Investment Trust Ltd.; architects: Guy Morgan & Partners; consulting engineers: Bylander, Waddell & Partners; quantity surveyors: Gardiner & Theobald.

6. Central Terminal buildings and main access tunnel, Heathrow – the start of twenty years of work at the airport. Client: British Airports Authority; architects: Frederick Gibberd & Partners; consulting engineers: Sir William Halcrow & Partners in association with G. H. Buckle & Partners & Ewbank & Partners; quantity surveyors: Rider Hunt & Partners, E. C. Harris & Partners, Franklin & Andrews. Multi-storey car parks: consulting engineers: L. G. Mouchel & Partners in association with Taylor Woodrow Construction Ltd.; architectural supervision: Frederick Gibberd & Part-

ners. Motor transport base:– architects: Clive Pascall & Peter Watson; consulting engineers: W. S. Atkins & Partners; quantity surveyors: C. E. Ball & Partners. Cargo tunnel:– consulting engineers: Sir William Halcrow & Partners.

7. Rangoon University buildings, for the Government of Burma Special Projects Implementing Board and carried out in association with United Burmese Engineers Ltd.; architect: Raglan Squire & Partners; consulting engineers (for Engineering College & Assembly Hall): Ove Arup & Partners.

8. H. M. Queen Elizabeth II inaugurates Calder Hall. Client: United Kingdom Atomic Energy Authority; engineers: Chief Engineers, Ministry of Works and Atomic Energy Authority.

Public Observation Platform in Gracechurch Street. Client: Midland Bank Ltd.; architects: Whinney, Son & Austen Hall; consulting engineers: Sir Frederick S. Snow & Partners; quantity surveyors: Reynolds & Young.

10. Lednock Dam, for the North of Scotland Hydro-Electric Board; consulting engineers: Sir M. Macdonald & Partners.

State visitor passes Esso House, Victoria Street – the first of three projects for The Land Securities Investment Trust Ltd. covering a quarter-mile frontage in Victoria Street. Architects: T. P. Bennett & Son; consulting engineers: Bylander, Waddell & Partners; quantity surveyors: Gardiner & Theobald.

11. Port Harcourt extensions, for the Nigerian Ports Authority; consulting engineers: Coode & Partners.

Opencast coal production, for the National Coal Board Opencast Executive.

12. Hinkley Point 'A' nuclear power station, designed and constructed for the Central Electricity Generating Board by British Nuclear Design Construction Ltd.

13. Nandi Airport, Fiji, for the Department of Civil Aviation, Commonwealth of Australia.

Ocean Terminal, for the Hong Kong & Kowloon Wharf & Godown Co. Ltd.; architects: Spence, Robinson, Prescott & Thornburrow; consulting engineers: S. E. Faber & Son; structural design: Taylor Woodrow and Phillips Consultants Ltd.; mechanical & electrical consultants: Thomas Anderson & Partners; quantity surveyors: Langdon & Every (Far East).

14. Arcon warehouses at Doha Deep Water Port, for the Government of Qatar.

15. Hotel Inter-Continental, Kabul, for Mailmah Pall Hotel Company. Consulting architects: Paton, Pitt & Associates; structural engineers: Phillips Consultants Ltd.; mechanical & electrical engineers: John Harvey & Partners; resource planning: Engineer Planning & Resources Ltd.

Shaft raising at Sizewell 580 MW nuclear power station designed and constructed for the Central Electricity Generating Board by British Nuclear Design & Construction Ltd.

17. Heron's Hill Development, Willowdale, Ontario, for Monarch Construction Ltd. Architects: Pentland, Baker and Polson; consulting engineers: Bradstock, Reicher & Partners Limited (structural); G. Granek and Associates (mechanical).

Swiftplan at Syon Park, for Gardening Centre Ltd.

20. Industrialised housing by Taylor Woodrow-Anglian Ltd. at Broadwater Farm, for the London Borough of Haringey; architect: C. E. Jacob Esq., A.R.I.B.A.; consulting engineers: Clark Nicholls & Marcel; quantity surveyors: Mercer & Miller, F/F.R.I.C.S.

22. Trans-Pennine Pipeline, for Imperial Chemical Industries Ltd.; consulting engineers: Pencol Engineering Consultants.

23. Sizewell nuclear power station, designed and constructed for the Central Electricity Generating Board by British Nuclear Design & Construction Ltd.

25. Glasgow Stock Exchange, for Scottish Metropolitan Property Co. Ltd.; architects: Baron Bercott & Associates; consulting engineers: W. V. Zinn & Associates; quantity surveyors: C. R. Wheeler & Partners.

University buildings in New Jersey for Rutgers, the State University; architects: Gruzen & Partners of New York; contractors: Blitman Construction Corporation.

26. Churchill Square, Brighton, Sussex – part of a comprehensive development being carried out by Myton Ltd. and the Standard Life Assurance Co. Ltd. in conjunction with Brighton Corporation. Architects: Russell Diplock Associates; planning consultant for Brighton Corporation: Sir Hugh Casson, A.R.A., R.D.I., M.A. (Cantab.), F.R.I.B.A. F.S.I.A.; consulting engineers: Phillips Consultants Ltd. quantity surveyors: Rider Hunt & Partners.

Conveyor tunnel, Holyhead, for Anglesey Aluminium Ltd.; engineers: Imperial Smelting/Kaiser Engineers.

27. East Lagoon Wharves, for the Port of Singapore Authority; consulting engineers: Sir Bruce White, Wolfe Barry & Partners.

Glass Container Manufacturing Plant, Selangor, Malaysia for Kuala Lumpur Glass Manufacturers Company, Sdn. Bhd., a subsidiary of Malaya Containers Berhad. Architects: James Ferrie & Partners; consulting engineers: Y. Wong, Sehu & Rakan.

28. St Katharine by the Tower – a Taylor Woodrow development in association with the Greater London Council. Architects: Renton Howard Wood Associates; consulting engineers: Ove Arup & Partners.

Taybol at Bournemouth, for Normandie Dredging, main contractor to Bournemouth Borough Council.

29. Terresearch on Morecambe Bay Sands, for the Water

Resources Board; consulting engineers: Sir Alexander Gibb & Partners.

Aluminium Smelter, Invergordon, Scotland for the British Aluminium Co. Ltd. being carried out by Taywood Wrightson Ltd. as managing contractor with responsibility for providing management, engineering, procurement and construction of the 100,000 tons per year plant.

30. New Midland Region Headquarters for the British Broadcasting Corporation at Edgbaston, Birmingham. Consulting architects: The John Madin Design Group in association with Mr R. A. Brown, Head of the Building Department, B.B.C.; consulting engineers (structural) Roy Bolsover & Associates; quantity surveyors: Messrs Ainsley.

Pilgrim Street, Newcastle, for the Corporation of Newcastle upon Tyne; architects: Robert Matthew, Johnson-Marshall & Partners in association with Mr George Kenyon, Dip. Arch. A.R.I.B.A., Dip. T.P., A.M.P.T.I., City Architect for Newcastle; consulting engineers: Ove Arup & Partners (structural), Steensen, Varming, Mulcahy & Partners (services); quantity surveyors: George Berry, F.R.I.C.S.

31. Hartlepool nuclear power station, being designed and constructed for the Central Electricity Generating Board by British Nuclear Design & Construction Ltd. Architects: Frederick Gibberd & Partners; consulting engineers: (for the seaward end of the circulating water outfall works) Merz & McLellan.

Index